Study Guide

Understanding Nutrition

ELEVENTH EDITION

Ellie Whitney

Sharon Rady Rolfes

Prepared by

Lori W. Turner
University of Alabama

THOMSON

WADSWORTH

Australia • Brazil • Canada • Mexico • Singapore • Spain • United Kingdom • United States

Printed in the United States of America

1 2 3 4 5 6 7 11 10 09 08 07

Printer: Thomson/West

ISBN-13: 978-0-495-11670-7
ISBN-10: 0-495-11670-X

Thomson Higher Education
10 Davis Drive
Belmont, CA 94002-3098
USA

For more information about our products, contact us at:
Thomson Learning Academic Resource Center
1-800-423-0563

For permission to use material from this text or product, submit a request online at
http://www.thomsonrights.com.
Any additional questions about permissions can be submitted by email to **thomsonrights@thomson.com.**

◎ Table of Contents ◎

⑥ What's in this Study Guide? ⑥

Thank you for purchasing the Study Guide for *Understanding Nutrition*, 11th edition! The exercises in this workbook are designed to help you prepare for examinations by reviewing key concepts and testing your recall of information presented in each textbook chapter. Answers for all questions are provided in an answer key at the end of each Study Guide section. The features of this Study Guide include:

- **Chapter Outlines** – Scan the outline to quickly review the topics covered within the chapter

- **Summing Up** – Fill in the blanks to complete the chapter summary

- **Chapter Study Questions** – Answer these discussion questions to practice explaining important processes and concepts in your own words

- **Chapter Glossaries** – Review the key terms and definitions from the chapter

- **Sample Test Questions** – Take this multiple-choice test to see how much you remember

- **Short Answer Questions** – Check your memory with these classifying and listing exercises

- **Problem Solving** – Practice application of chapter concepts through nutrition calculations and other word problems

- **Figure Identification** – Label anatomical diagrams and chemical structures

- **Crossword Puzzles** – Have some fun while reviewing key terms

⑤ Chapter 1 – An Overview of Nutrition ⑤

Chapter Outline

I. Food Choices
   A. Personal Preference
   B. Habit
   C. Ethnic Heritage or Tradition
   D. Social Interactions
   E. Availability, Convenience, and Economy
   F. Positive and Negative Associations
   G. Emotional Comfort
   H. Values
   I. Body Weight and Image
   J. Nutrition and Health Benefits

II. The Nutrients
   A. Nutrients in Foods and in the Body
     1. Composition of Foods
     2. Composition of the Body
     3. Chemical Composition of Nutrients
     4. Essential Nutrients
   B. The Energy-Yielding Nutrients: Carbohydrate, Fat, and Protein
     1. Energy Measured in kCalories
     2. Energy from Foods
     3. Energy in the Body
     4. Other Roles of Energy-Yielding Nutrients
   C. The Vitamins
   D. The Minerals
   E. Water

III. The Science of Nutrition
   A. Conducting Research
     1. Controls
     2. Sample Size
     3. Placebos
     4. Double Blind
   B. Analyzing Research Findings
     1. Correlations and Causes
     2. Cautious Conclusions
   C. Publishing Research

IV. Dietary Reference Intakes
   A. Establishing Nutrient Recommendations
     1. Estimated Average Requirements (EAR)
     2. Recommended Dietary Allowances (RDA)
     3. Adequate Intakes (AI)
     4. Tolerable Upper Intake Levels (UL)
   B. Establishing Energy Recommendations
     1. Estimated Energy Requirement (EER)
     2. Acceptable Macronutrient Distribution Ranges (AMDR)
   C. Using Nutrient Recommendations
   D. Comparing Nutrient Recommendations

V. Nutrition Assessment
   A. Nutrition Assessment of Individuals
     1. Historical Information
     2. Anthropometric Data
     3. Physical Examinations
     4. Laboratory Tests
     5. Iron, for Example
   B. Nutrition Assessment of Populations
     1. National Nutrition Surveys
     2. National Health Goals
     3. National Trends

VI. Diet and Health
   A. Chronic Diseases
   B. Risk Factors for Chronic Diseases
     1. Risk Factors Persist
     2. Risk Factors Cluster
     3. Risk Factors in Perspective

VII. Nutrition Information and Misinformation—On the Net and in the News
   A. Nutrition on the Net
   B. Nutrition in the News
   C. Identifying Nutrition Experts
     1. Physicians and Other Health Care Professionals
     2. Registered Dietitians
     3. Other Dietary Employees
   D. Identifying Fake Credentials
   E. Red Flags of Nutrition Quackery

Summing Up

People eat for many reasons other than to obtain the (1)_____ they need. Reasons people select the foods they do include: personal preference, habit, ethnic heritage or tradition, (2)_____, availability, convenience, economy, positive and negative associations, (3)_____, values, image, and nutrition.

The human body is made almost entirely of (4)_____ derived from food. The six classes of nutrients are: carbohydrate, (5)_____ (lipid), protein, (6)_____ (fat soluble and water soluble), minerals, and (7)_____. The first four are (8)_____; the first three provide (9)_____ in a form the body can use. (Alcohol also releases energy, but is not a (10)_____; it isn't used for (11)_____, maintenance, or repair of body tissues.) The (12)_____ nutrients must be obtained preformed from food. Food energy is measured in (13)_____, units of heat, or kilojoules, units of (14)_____.

The research process involves testing of the (15)_____. When researchers divide the subjects into two groups, one is the (16)_____ group and one is the control group. The control group may receive a pill that contains no medicine or treatment to control for the (17)_____ effect. When analyzing studies, a correlation between variables may be found, but one may not be the (18)_____ of the other. Before it is published, research must go through (19)_____ review. Other scientists need to confirm or disprove the findings through (20)_____.

The DRI—(21)_____ Reference Intakes—represent a set of values for the dietary nutrient intakes of healthy people. The four values include: (22)_____ Average Requirements, (23)_____ Dietary Allowances, (24)_____ Intakes, and (25)_____ Upper Intake Levels. The (26)_____ Energy Requirement (EER) represents the average dietary energy intake that will maintain energy balance in a person who has a healthy body weight and level of physical activity. The (27)_____ Macronutrient Distribution Ranges (AMDR) represent the composition of a diet that provides adequate energy and nutrients and reduces the risk of chronic diseases.

A registered dietitian will use four methods to provide nutrition (28)_____ of individuals. They are historical information, (29)_____ data, physical examinations, and (30)_____ tests. Given too much, too little, or an imbalance of nutrients, the body becomes (31)_____. The sequence of events in a deficiency begins with poor nutrient (32)_____, or inefficient absorption or use of the nutrient, then progresses to internal abnormalities, and finally manifests itself in externally observable (33)_____. To provide nutrition assessments of populations, researchers use (34)_____ Nutrition Surveys that examine dietary quality, body composition, physical activity and other variables.

Factors that increase or reduce the risk of developing disease are called (35)_____ _____. Risk factors tend to (36)_____ over time and cluster. The most predominant risk factor contributing to disease is (37)_____ use, followed closely by (38)_____ and activity patterns.

Chapter Study Questions

1. Give several reasons (and examples) why people make the food choices that they do.

2. What is a nutrient? Name the six classes of nutrients found in foods. What is an essential nutrient?

3. Which nutrients are inorganic and which are organic? Discuss the significance of that distinction.

4. Which nutrients yield energy and how much energy do they yield per gram? How is energy measured?

5. Describe how alcohol resembles nutrients. Why is alcohol not considered a nutrient?

6. What is the science of nutrition? Describe the types of research studies and methods used in acquiring nutrition information.

7. Explain how variables might be correlational but not causal.

8. What are the DRI? Who develops the DRI? To whom do they apply? How are they used? In your description, identify the four categories of DRI and indicate how they are related.

9. What judgment factors are involved in setting the energy and nutrient intake recommendations?

10. What happens when people either get too little or too much energy or nutrients? Define malnutrition, undernutrition, and overnutrition. Describe the four methods used to detect energy and nutrient deficiencies and excesses.

11. What methods are used in nutrition surveys? What kinds of information can these surveys provide?

12. Describe risk factors and their relationships to disease.

Chapter Glossary

- **Acceptable Macronutrient Distribution Ranges (AMDR):** ranges of intakes for the energy nutrients that provide adequate energy and nutrients and reduce the risk of chronic diseases.
- **Adequate Intake (AI):** the average daily amount of a nutrient that appears sufficient to maintain a specified criterion; a value used as a guide for nutrient intake when an RDA cannot be determined.
- **anthropometric:** relating to measurement of the physical characteristics of the body, such as height and weight.
- **calories:** units by which energy is measured. Food energy is measured in kilocalories (1000 calories equal 1 kilocalorie), abbreviated kcalories or kcal. One kcalorie is the amount of heat necessary to raise the temperature of 1 kilogram (kg) of water 1°C. The scientific use of the term kcalorie is the same as the popular use of the term calorie.
- **chronic diseases:** diseases characterized by a slow progression and long duration. Examples include heart disease, cancer, and diabetes.
- **covert:** hidden, as if under covers.
- **deficient:** the amount of a nutrient below which almost all healthy people can be expected, over time, to experience deficiency symptoms.
- **diet:** the foods and beverages a person eats and drinks.
- **Dietary Reference Intakes (DRI):** a set of nutrient intake values for healthy people in the United States and Canada. These values are used for planning and assessing diets and include Estimated Average Requirements (EAR), Recommended Dietary Allowances (RDA), Adequate Intakes (AI), and Tolerable Upper Intake Levels (UL).
- **energy:** the capacity to do work. The energy in food is chemical energy. The body can convert this chemical energy to mechanical, electrical, or heat energy.
- **energy density:** a measure of the energy a food provides relative to the amount of food (kcalories per gram).
- **energy-yielding nutrients:** the nutrients that break down to yield energy the body can use—carbohydrate, fat, and protein.

- **essential nutrients:** nutrients a person must obtain from food because the body cannot make them for itself in sufficient quantity to meet physiological needs; also called indispensable nutrients. About 40 nutrients are currently known to be essential for human beings.
- **Estimated Average Requirement (EAR):** the average daily amount of a nutrient that will maintain a specific biochemical or physiological function in half the healthy people of a given age and gender group.
- **Estimated Energy Requirement (EER):** the average dietary energy intake that maintains energy balance and good health in a person of a given age, gender, weight, height, and level of physical activity.
- **foods:** products derived from plants or animals that can be taken into the body to yield energy and nutrients for the maintenance of life and the growth and repair of tissues.
- **functional foods:** foods that contain physiologically active compounds that provide health benefits beyond their nutrient contributions; sometimes called designer foods or nutraceuticals.
- **genome:** the full complement of genetic material (DNA) in the chromosomes of a cell. In human beings, the genome consists of 46 chromosomes. The study of genomes is called genomics.
- *Healthy People*: a national public health initiative under the jurisdiction of the U.S. Department of Health and Human Services (DHHS) that identifies the most significant preventable threats to health and focuses efforts toward eliminating them.
- **inorganic:** not containing carbon or pertaining to living things.
- **malnutrition:** any condition caused by excess or deficient food energy or nutrient intake or by an imbalance of nutrients.
- **minerals:** inorganic elements. Some minerals are essential nutrients required in small amounts by the body for health.
- **nutrients:** chemical substances obtained from food and used in the body to provide energy, structural materials, and regulating agents to support growth, maintenance, and repair of the body's tissues. Nutrients may also reduce the risks of some diseases.
- **nutrition:** the science of foods and the nutrients and other substances they contain, and of their actions within the body (including ingestion, digestion, absorption, transport, metabolism, and excretion). A broader definition includes the social, economic, cultural, and psychological implications of food and eating.
- **nutritional genomics:** the science of how nutrients affect the activities of genes (nutrigenomics) and how genes affect the interactions between diet and disease (nutrigenetics).
- **nutrition assessment:** a comprehensive analysis of a person's nutrition status that uses health, socioeconomic, drug, and diet histories; anthropometric measurements; physical examinations; and laboratory tests.
- **organic:** in chemistry, a substance or molecule containing carbon-carbon bonds or carbon-hydrogen bonds. This definition excludes coal, diamonds, and a few carbon-containing compounds that contain only a single carbon and no hydrogen, such as carbon dioxide (CO_2), calcium carbonate ($CaCO_3$), magnesium carbonate ($MgCO_3$), and sodium cyanide (NaCN).
- **overnutrition:** excess energy or nutrients.
- **overt:** out in the open and easy to observe.
- **phytochemicals:** nonnutrient compounds found in plant-derived foods that have biological activity in the body.
- **primary deficiency:** a nutrient deficiency caused by inadequate dietary intake of a nutrient.
- **Recommended Dietary Allowance (RDA):** the average daily amount of a nutrient considered adequate to meet the known nutrient needs of practically all healthy people; a goal for dietary intake by individuals.
- **requirement:** the lowest continuing intake of a nutrient that will maintain a specified criterion of adequacy.
- **risk factor:** a condition or behavior associated with an elevated frequency of a disease but not proved to be causal. Leading risk factors for chronic diseases include obesity, cigarette smoking, high blood pressure, high blood cholesterol, physical inactivity, and a diet high in saturated fats and low in vegetables, fruits, and whole grains.
- **secondary deficiency:** a nutrient deficiency caused by something other than an inadequate intake such as a disease condition or drug interaction that reduces absorption, accelerates use, hastens excretion, or destroys the nutrient.
- **subclinical deficiency:** a deficiency in the early stages, before the outward signs have appeared.
- **Tolerable Upper Intake Level (UL):** the maximum daily amount of a nutrient that appears safe for most healthy people and beyond which there is an increased risk of adverse health effects.

- **undernutrition:** deficient energy or nutrients.
- **vitamins:** organic, essential nutrients required in small amounts by the body for health.

Sample Test Questions

Select the best answer for each question.

1. When people eat foods associated with their families and geographical location, their eating patterns are being influenced by:
   - a. personal preference.
   - b. habit.
   - c. ethnic heritage or tradition.
   - d. physical appearance.
   - e. nutrition.

2. People who eat foods to relieve boredom or calm anxiety are eating for reasons of:
   - a. availability.
   - b. nutrition.
   - c. values.
   - d. emotional comfort.
   - e. economy.

3. The science of nutrition is the study of:
   - a. how to prepare delicious foods that will improve a person's health.
   - b. nutrient ingestion, digestion, absorption, and transport.
   - c. nutrient metabolism, interaction, storage, and excretion.
   - d. a and b
   - e. b and c

4. A complete chemical analysis of your body would show that it is composed mostly of:
   - a. proteins.
   - b. carbohydrates.
   - c. fats.
   - d. water.
   - e. minerals.

5. An organic compound is:
   - a. a compound that contains carbon and hydrogen atoms.
   - b. any substance found in living organisms.
   - c. a compound that contains oxygen atoms.
   - d. found only in foods grown under special conditions.
   - e. superior.

6. Among these classes of nutrients, which one is not organic?
   - a. carbohydrate
   - b. fat
   - c. protein
   - d. vitamins
   - e. minerals

7. Which does not yield energy for human use?
   - a. carbohydrates
   - b. vitamins
   - c. fats
   - d. proteins

8. When energy nutrients are metabolized:
   - a. the arrangement of atoms remains unaltered.
   - b. energy is released.
   - c. the bonds between the nutrients' atoms break.
   - d. a and b
   - e. b and c

9. A kcalorie is a:
    a. gram of fat.
    b. unit in which energy is measured.
    c. heating device.
    d. term used to describe the amount of sugar and fat in foods.

10. Any food can be "fattening" if you eat too much of it, even protein-rich food.
    a. True   b. False

11. A carbohydrate-rich food like bread:
    a. contains a mixture of the three energy nutrients.
    b. contains no protein.
    c. contains no protein or fat.
    d. cannot be properly classified on the basis of nutrient content.

12. One gram of alcohol contains:
    a. 4 kcal.   c. 9 kcal.
    b. 7 kcal.   d. 0 kcal.

13. Which of the following are fat soluble?
    1. vitamin A   4. vitamin D
    2. B vitamins   5. vitamin E
    3. vitamin C   6. vitamin K

    a. 1, 4, 5, 6   d. 2, 3
    b. 2, 3, 5   e. 2, 5, 6
    c. 1, 4

14. Minerals are:
    a. energy providers.   c. large.
    b. inorganic elements.   d. organic elements.

15. Water:
    a. is organic.   c. is indispensable.
    b. gives us energy.   d. is not a nutrient.

16. This type of research study observes how much and what kinds of foods a group of people eat:
    a. human intervention.   c. case-control.
    b. animal.   d. epidemiological.

17. The _____ are a set of four nutrient intake values that can be used to plan and evaluate diets for healthy people.
    a. RDA   c. TUIL
    b. AI   d. DRI

18. The dietary recommendations apply to:
    a. average daily intakes.   d. a and b
    b. healthy people.   e. a and c
    c. all people.

19. Historical information, physical examination, laboratory tests, and anthropometric measures are:
    a. steps in the scientific method.
    b. methods used in nutrition assessment.
    c. procedures used to determine nutrient content of a diet.
    d. obsolete procedures used by nutritionists years ago.

20. A deficiency caused by an inadequate intake of a nutrient is a _____ deficiency.
    a. primary
    b. secondary
    c. dire
    d. clinical
    e. subclinical

21. A technique to detect nutrient deficiencies by taking height and weight measurements is part of the nutrition assessment component known as:
    a. diet history.
    b. anthropometrics.
    c. physical examination.
    d. biochemical tests.

22. Factors associated with elevated frequency of a disease but not proved to be causal are:
    a. medical factors.
    b. individual interventions.
    c. risk factors.
    d. disease clusters.

23. A research study often involves testing a(n):
    a. anecdote.
    b. sample size.
    c. validity.
    d. hypothesis.

24. A group that is not given the treatment is called the
    a. placebo.
    b. control group.
    c. experimental group.
    d. randomized group.

25. What study design is used to control for the placebo effect?
    a. sample size design
    b. blind design
    c. positive correlational design
    d. variable design

26. A correlation between variables indicates that the:
    a. variables are associated.
    b. variables are causative.
    c. variables have a causative mechanism.
    d. all of the above

27. What process occurs when scientists evaluate studies for publication potential?
    a. validity assessment
    b. replication
    c. experimental review
    d. peer review

28. The average dietary energy intake required to maintain energy balance in a person who has a healthy body weight and level of physical activity is the:
    a. Tolerable Upper Intake Level (UL).
    b. Recommended Dietary Allowance (RDA).
    c. Estimated Energy Requirement (EER).
    d. Adequate Intake (AI).

29. According to the Acceptable Macronutrient Distribution Ranges (AMDR), the appropriate percentages of kcalories from carbohydrate, protein and fat are:
    a. 35-45% carbohydrate, 20-35% fat, and 20-45% protein.
    b. 45-65% carbohydrate, 20-35% fat, and 10-35% protein.
    c. 45-65% carbohydrate, 10-25% fat, and 20-45% protein.
    d. 50-75% carbohydrate, 5-20% fat, and 10-35% protein.

30. Which of the following is true about recommendations?
    a. They are minimal requirements.
    b. They are optimal intakes for all individuals.
    c. They can target most people.
    d. They can account for individual variations in nutrient needs.

Short Answer Questions

1. The six classes of nutrients are:

    a.   d.

    b.   e.

    c.   f.

2. Which nutrients are organic?

    a.   c.

    b.   d.

3. Which nutrients are inorganic?

    a.   b.

4. Which nutrients provide energy?

    a.   c.

    b.

5. 1 gram carbohydrate   = _____ kcalories

    1 gram fat   = _____ kcalories

    1 gram protein   = _____ kcalories

    1 gram alcohol   = _____ kcalories

6. ½ cup vegetables   = _____ mL or _____ grams

    ½ cup juice or milk   = _____ mL or _____ grams

7. The water-soluble vitamins are:

    a.   b.

8. The fat-soluble vitamins are:

    a.   c.

    b.   d.

9. Name and describe the methods used in nutrition assessment.

    a.

    b.

    c.

    d.

Problem Solving

1. How many grams of fat were consumed if a person received 405 kcalories from fat in a day?

2. How many kcalories are in 13 grams of carbohydrate?

3. How many total kcalories are in 10 grams of carbohydrate, 4 grams of protein, 6 grams of fat, and 7 grams of alcohol?

4. If a food product contains 20 grams of protein, how many kcalories will be provided from protein?

5. If a food item contains 10 grams of protein and another food item contains 10 grams of carbohydrate, how many kcalories are provided from each and which item is more fattening?

6. If an alcoholic beverage contains 1 gram of ethanol and 10 grams of carbohydrate, how many kcalories does it provide?

7. A person consumes 100 grams of fat, 50 grams of carbohydrate, 30 grams of protein, 20 milligrams of thiamin and 250 milligrams of calcium. How much energy did that person consume?

8. A woman consumes 500 grams of carbohydrate, 30 grams of protein, and 75 grams of fat in one day. How many total kcalories did she consume, and how many and what percentage of kcalories were from carbohydrate, protein and fat?

9. A person consumes 400 grams of carbohydrate, 150 grams of protein, 300 grams of fat and 5 grams of ethanol in one day. How many total kcalories were consumed, and how many and what percent of kcalories were from carbohydrate, protein, fat, and alcohol?

10. Meal A provides 250 grams of protein and 40 grams of fat. Meal B provides 250 grams of carbohydrate and 40 grams of fat. How many kcalories are provided by each meal?

10

Crossword Puzzle

Complete this crossword puzzle by Mary A. Wyandt, Ph.D., CHES.

	Across		Down
1.	an experiment in which the subjects do not know whether they are members of the experimental or the control group	2.	the healing effect that faith in medicine, even inert medicine, often has
5.	repeating an experiment and getting the same results	3.	a group of individuals similar in all possible respects to the group being experimented on except for the experimental treatment
6.	the simultaneous increase, decrease, or change of two variables	4.	a factor that changes
7.	a group of individuals similar in all possible respects to the control group except for the treatment	8.	having the quality of being founded on fact or evidence
10.	a process of choosing the members of the experimental and control groups without bias	9.	people or animals participating in a research project

⑥ Chapter 1 Answer Key ⑥

Summing Up

1. nutrition
2. social interactions
3. emotional comfort
4. nutrients
5. fat
6. vitamins
7. water
8. organic
9. energy
10. nutrient
11. growth
12. essential
13. kcalories
14. work
15. hypothesis
16. experimental
17. placebo
18. cause
19. peer
20. replication
21. Dietary
22. Estimated
23. Recommended
24. Adequate
25. Tolerable
26. Estimated
27. Acceptable
28. assessment
29. anthropometric
30. laboratory
31. malnourished
32. intake
33. symptoms
34. National
35. risk factors
36. accumulate
37. tobacco
38. diet.

Chapter Study Questions

1. Personal preference, habit, ethnic heritage or tradition, social interactions, availability, convenience, economy, positive and negative associations, emotional comfort, values, image, nutrition.
2. A substance obtained from food and used in the body to promote growth, maintenance, and repair. Carbohydrate, fat, protein, vitamins, minerals, and water. Essential: must be obtained from an outside source.
3. Organic nutrients: carbohydrates, fats, proteins, vitamins. Inorganic nutrients: minerals, water. Organic means all carbon compounds, not necessarily living things.
4. Energy-yielding nutrients: carbohydrate (4 kcal/g), fat (9 kcal/g), protein (4 kcal/g). Measured in calories or kilocalories.
5. Alcohol yields energy (7 kcal per gram) when metabolized, but alcohol is not considered a nutrient because it does not support the growth, maintenance, or repair of the body.
6. The science of nutrition is the study of the nutrients in foods and the body's handling of those nutrients. Epidemiological studies, case-control studies, animal studies, and human intervention (or clinical) trials are all used to acquire information.
7. Correlational variables are associated with each other; causal variables require a mechanism and indicate that one causes the other.
8. The DRI are the Dietary Reference Intakes. They are a set of four nutrient intake values that can be used to plan and evaluate diets for healthy people. They are developed by the DRI Committee. Members of the committee are selected from the Food and Nutrition Board of the Institute of Medicine, the National Academy of Sciences, and Health Canada. The four categories include the Estimated Average Requirement (defines the amount of a nutrient that supports a specific function in the body for half of the population); the Recommended Dietary Allowance (uses Estimated Average Requirement to establish a goal for dietary intake that will meet the needs of almost all healthy people); an Adequate Intake (serve a similar purpose when an RDA cannot be determined); and Tolerable Upper Intake Level (establishes the highest amount that appears safe for regular consumption).
9. How much of a nutrient a person needs, which is determined by studying deficiency states, nutrient stores, and depletion, and by measuring the body's intake and excretion of the nutrient; that different individuals have different requirements; at what dividing line the bulk of the population is covered.
10. They get sick and show signs of deficiencies. Malnutrition is poor nutrition status, undernutrition is underconsumption of food energy or nutrients severe enough to cause disease or increased susceptibility to disease, and overnutrition is overconsumption of food energy or nutrients severe enough to cause disease or to cause increased susceptibility to disease. They are detected through nutrition assessment techniques (anthropometric measures, lab tests, physical findings, and diet history).
11. Administering questionnaires, conducting interviews, collecting anthropometric measurements, and physical examinations on groups of people are methods used in nutrition surveys. Food consumption surveys and

nutrition status surveys can provide information regarding amounts and kinds of foods people consume as well as evaluate people's nutrition status.

12. Factors that increase the risk of developing chronic diseases are called risk factors. A strong association between a risk factor and a disease means that when the factor is present, the likelihood of developing the disease increases.

Sample Test Questions

1. c (p. 4)	9. b (p. 7)	17. d (p. 16)	25. b (p. 12-13)
2. d (p. 4)	10. a (p. 10)	18. d (p. 16, 19)	26. a (p. 14-15)
3. e (p. 11)	11. a (p. 9)	19. b (p. 20)	27. d (p. 15)
4. d (p. 6)	12. b (p. 8)	20. a (p. 22)	28. c (p. 18)
5. a (p. 7)	13. a (p. 10n)	21. b (p. 21)	29. b (p. 18)
6. e (p. 7)	14. b (p. 7)	22. c (p. 24)	30. c (p. 19)
7. b (p. 7)	15. c (p. 11)	23. d (p. 11)	
8. e (p. 9-10)	16. d (p. 13)	24. b (p. 12)	

Short Answer Questions

1. carbohydrate; fat; protein; vitamins; minerals; water
2. carbohydrate; fat; protein; vitamins
3. minerals; water
4. carbohydrate; fat; protein
5. carbohydrate = 4; fat = 9; protein = 4; alcohol = 7
6. vegetables = 120 mL or 100 g; liquid = 120 mL or 100 g
7. B vitamins; vitamin C
8. vitamin A; vitamin D; vitamin E; vitamin K
9. data on diet – record foods eaten over a period of time; physical examination – inspect body parts (hair, eyes, skin); laboratory tests – analyze body samples (blood, urine); anthropometric measures – measure height, weight, body parts

Problem Solving

1. 45 g (405 kcal divided by 9 kcal/g fat = 45 g fat)

2. 52 kcal (13 g carb multiplied by 4 kcal/g carb = 52 kcal)

3. 159 kcal (10 g carb x 4 kcal/g + 4 g pro x 4 kcal/g + 6 g fat x 9 kcal/g + 7 g alc x 7 kcal/g = 40 + 16 + 54 + 49 = 159 kcal)

4. 20 g x 4 = 80 kcalories from protein

5. 10 g x 4 = 40 kcalories from protein
   10 g x 4 = 40 kcalories from carbohydrate
   They each provide 40 kcalories and they are equally fattening.

6. 1 g ethanol x 7 = 7 kcalories from ethanol
   10 g carbohydrate x 4 = 10 kcalories from carbohydrate
   7 + 40 = 47 kcalories

7. 100 g fat x 9 = 900 kcalories from fat
   50 g carbohydrate x 4 = 200 kcalories from carbohydrate
   30 g protein x 4 = 120 kcalories from protein
   20 mg thiamine = 0 kcalories
   250 mg calcium = 0 kcalories
   900 + 200 + 120 = 1220 total kcalories

8. 500 g carbohydrate x 4 = 2000 kcal
   30 g protein x 4 = 120 kcal
   75 g fat x 9 = 675 kcal
   2000 + 120 + 675 = 2795 total kcalories
   2000/2795 = 72% kcalories from carbohydrate
   120/2795 = 4% kcalories from protein
   675/2795 = 24% kcalories from fat

9. 400 g carbohydrate x 4 = 1600 kcalories
   150 g protein x 4 = 600 kcalories
   300 g fat x 9 = 2700 kcalories
   5 g ethanol x 7 = 35 kcalories
   1600 + 600 + 2700 + 35 = 4935 total kcalories
   1600/4935 = 32% kcalories from carbohydrate
   600/4935 = 12% kcalories from protein
   2700/4935 = 55% kcalories from fat
   35/4935 = 1% kcalories from ethanol

10. Meal A: 250 g protein x 4 = 1000 kcal; 40 g fat x 9 = 360 kcal; 1000 + 360 = 1360 kcalories
    Meal B: 250 g carbohydrate x 4 = 1000 kcal; 40 g fat x 9 = 360 kcal; 1000 + 360 = 1360 kcalories
    They have the same number of kcalories.

Crossword Puzzle

1. blind experiment
2. placebo effect
3. control group
4. variable

5. replication
6. correlation
7. experimental group
8. validity

9. subjects
10. randomization

⑥ Chapter 2 – Planning a Healthy Diet ⑨

Chapter Outline

I. Principles and Guidelines
    A. Diet-Planning Principles
        1. Adequacy
        2. Balance
        3. kCalorie (Energy) Control
        4. Nutrient Density
        5. Moderation
        6. Variety
    B. Dietary Guidelines for Americans
II. Diet-Planning Guides
    A. USDA Food Guide
        1. Recommended Amounts
        2. Notable Nutrients
        3. Nutrient Density
        4. Discretionary kCalorie Allowance
        5. Serving Equivalents
        6. Mixtures of Foods
        7. Vegetarian Food Guide
        8. Ethnic Food Choices
        9. MyPyramid—Steps to a Healthier You
    B. Exchange Lists
    C. Putting the Plan into Action
    D. From Guidelines to Groceries
        1. Grains
        2. Vegetables
        3. Fruit
        4. Meat, Fish, and Poultry

        5. Milk
III. Food Labels
    A. The Ingredient List
    B. Serving Sizes
    C. Nutrition Facts
    D. The Daily Values
    E. Nutrient Claims
    F. Health Claims
    G. Structure-Function Claims
    H. Consumer Education
IV. Vegetarian Diets
    A. Health Benefits of Vegetarian Diets
        1. Weight Control
        2. Blood Pressure
        3. Heart Disease
        4. Cancer
        5. Other Diseases
    B. Vegetarian Diet Planning
        1. Protein
        2. Iron
        3. Zinc
        4. Calcium
        5. Vitamin B12
        6. Vitamin D
        7. Omega-3 Fatty Acids
    C. Healthy Food Choices

Summing Up

Six concepts to keep in mind when planning a nutritious diet are: (1)_____,

(2)_____, (3)_____ control, nutrient density, moderation, and

(4)_____. The *Dietary Guidelines* encourage people to eat a (5)_____ of foods

to get the (6)_____ needed to support good health and the energy appropriate to maintain a

healthy (7)_____. Food group plans and (8)_____ systems use these concepts.

Food group plans enable people to build diets from clusters of foods that are similar in

(9)_____ content. The *2005 Dietary* (10)_____ encourage consumers to adopt

a (11)_____ eating plan. The USDA's Food Guide assigns foods to (12)_____

major groups and recommends daily amounts of foods from each group to meet (13)_____ needs.

The USDA Food Guide encourages greater consumption from certain food groups to provide the nutrients most

often (14)_____. A foundation for a healthy diet emphasizes (15)_____-

_____ options. The difference between kcalories needed to supply nutrients and those needed for energy is the (16)_____ kcalorie allowance. MyPyramid helps consumers select the kinds and

(17)_____ of foods they should eat each day.

Exchange systems list foods according to the number of kcalories and the amount of

(18)_____, (19)_____, and (20)_____ each contains. Each

list defines specific portion (21)_____ for individual foods. Exchange systems are useful for

monitoring (22)_____ intakes.

When grocery shopping, it is wise to choose (23)_____-_____ breads

and cereals, fresh (24)_____ and fruits and moderate amounts of lean meats and

(25)_____ products. Current food labels provide consumers with useful information about the

foods they eat, and how individual foods fit into their daily (26)_____. The ingredient list

includes all ingredients in (27)_____ order of predominance by weight.

(28)_____ sizes must be identified on labels. Food labels must also include the amount of a

nutrient in a product as a (29)_____ of its Daily Value. The FDA has developed several programs

to (30)_____ consumers about labels.

Chapter Study Questions

1. Name the diet-planning principles and briefly describe how each principle helps in diet planning.

2. What recommendations appear in the *Dietary Guidelines for Americans*?

3. Name the five food groups in the USDA Food Guide and identify several foods typical of each group. Explain how this Guide groups foods, and how the Guide incorporates concepts of nutrient density and kcalorie control.

4. Review the *Dietary Guidelines*. What types of grocery selections would you make to achieve those recommendations?

5. Define the terms *notable nutrients, nutrient density,* and *discretionary kcalorie allowance.*

6. Describe the purpose of MyPyramid.

7. Discuss how exchange lists are used in diet planning.

8. What information can you expect to find on a food label? How can this information help you choose between two similar products?

9. What are the Daily Values? How can they help you meet health recommendations?

10. List and define the three types of claims that may be on a food product.

Chapter Glossary

- **adequacy (dietary):** providing all the essential nutrients, fiber, and energy in amounts sufficient to maintain health.
- **balance (dietary):** providing foods in proportion to each other and in proportion to the body's needs.
- **Daily Values (DV):** reference values developed by the FDA specifically for use on food labels.
- **discretionary kcalorie allowance:** the kcalories remaining in a person's energy allowance after consuming enough nutrient-dense foods to meet all nutrient needs for a day.
- **empty-kcalorie foods:** a popular term used to denote foods that contribute energy but lack protein, vitamins, and minerals.
- **enriched:** the addition to a food of nutrients that were lost during processing so that the food will meet a specified standard.
- **exchange lists:** diet-planning tools that organize foods by their proportions of carbohydrate, fat, and protein. Foods on any single list can be used interchangeably.
- **food group plans:** diet-planning tools that sort foods into groups based on nutrient content and then specify that people should eat certain amounts of foods from each group.
- **food substitutes:** foods that are designed to replace other foods.
- **fortified:** the addition to a food of nutrients that were either not originally present or present in insignificant amounts. Fortification can be used to correct or prevent a widespread nutrient deficiency or to balance the total nutrient profile of a food.
- **health claims:** statements that characterize the relationship between a nutrient or other substance in a food and a disease or health-related condition.

- **imitation foods:** foods that substitute for and resemble another food, but are nutritionally inferior to it with respect to vitamin, mineral, or protein content. If the substitute is not inferior to the food it resembles and if its name provides an accurate description of the product, it need not be labeled "imitation."
- **kcalorie (energy) control:** management of food energy intake.
- **legumes:** plants of the bean and pea family, with seeds that are rich in protein compared with other plant-derived foods.
- **moderation (dietary):** providing enough but not too much of a substance.
- **nutrient claims:** statements that characterize the quantity of a nutrient in a food.
- **nutrient density:** a measure of the nutrients a food provides relative to the energy it provides. The more nutrients and the fewer kcalories, the higher the nutrient density.
- **processed foods:** foods that have been treated to change their physical, chemical, microbiological, or sensory properties.
- **refined:** the process by which the coarse parts of a food are removed. When wheat is refined into flour, the bran, germ, and husk are removed, leaving only the endosperm.
- **structure-function claims:** statements that characterize the relationship between a nutrient or other substance in a food and its role in the body.
- **textured vegetable protein:** processed soybean protein used in vegetarian products such as soy burgers.
- **variety (dietary):** eating a wide selection of foods within and among the major food groups.
- **whole grain:** a grain milled in its entirety (all but the husk), not refined.

Sample Test Questions

Select the best answer for each question.

1. Diet planning principles include:
   a. adequacy, B vitamins, carbohydrates, moderation, nutrient density, and variety.
   b. abundance, balance, carbohydrates, meals, nutrients, and vegetables.
   c. adequacy, balance, kcalorie control, nutrient density, moderation, and variety.
   d. abundance, B vitamins, kcalorie control, milk, moderation, and vegetables.

2. Selecting foods that deliver the most nutrients for the least food energy is applying the concept of:
   a. moderation.
   b. abundance.
   c. variety.
   d. adequacy.
   e. nutrient density.

3. Which of the following is the most nutrient-dense food relative to calcium content?
   a. whole milk
   b. non-fat milk
   c. low-fat milk
   d. cheddar cheese

4. Foods that are low in nutrient density are:
   a. empty-kcalorie foods.
   b. raw vegetables.
   c. those that deliver protein, vitamins and minerals with little kcalories.
   d. low in fat and sugar.

5. Important topics that are part of the *Dietary Guidelines for Americans* include:
   a. choosing nutrient-dense foods.
   b. maintaining a healthy body weight.
   c. engaging in regular physical activity.
   d. all of the above

6. An adult following the Food Guide should consume a variety of vegetables from all _____ subgroups of vegetables each week.
   a. one
   b. two
   c. three
   d. four
   e. five

7. Which of the following is descriptive of the USDA Food Guide?
   a. a figure designed to assist the average consumer in the use of the Food Exchange System
   b. an education tool for teaching nutrition to children that consists of food blocks that require stacking in a specific order
   c. a system that assigns foods to five major groups and recommends daily amounts of foods from each group
   d. a system of specialized containers of several different sizes which allows for better storage and preservation of perishable food items

8. The Food Guide consists of the following groups:
   a. fruits, vegetables, grains, meat and legumes, milk.
   b. fruits, vegetables, grains, protein, oils.
   c. fruits and vegetables, breads, meat, fat, carbohydrate.
   d. fruits and vegetables, breads, meats, legumes, dairy.

9. Exchange lists group foods according to:
   1. protein, fat, and carbohydrate content.
   2. kcaloric content.
   3. vitamin content.
   4. mineral content.

   a. 1 only
   b. 2 only

   c. 3 and 4
   d. 1, 2, 3, and 4

10. The USDA Food Guide sorts the vegetable group into 5 subgroups for the following reason:
    a. the former plan was too simple to understand.
    b. people were consuming too many legumes.
    c. the vegetable group is widely enjoyed.
    d. some vegetables are especially good sources of certain nutrients.

11. The term *notable nutrients* refers to:
    a. nutrients that are needed in greater amounts than other nutrients.
    b. nutrients that have more important body functions than other nutrients.
    c. nutrients that are contained in empty-kcalorie foods.
    d. nutrients most often lacking in the diets of Americans.

12. The difference between kcalories needed to supply nutrients and those needed for energy is:
    a. nutrient density kcalories.
    b. notable nutrient kcalories.

    c. discretionary kcalories.
    d. empty kcalories.

13. An educational tool that illustrates the concepts of the *Dietary Guidelines* and USDA Food Guide is called:
    a. the Exchange Lists.
    b. the *Healthy People 2010* objectives.

    c. MyPyramid.
    d. the Sample Diet Plan.

14. Which of the following is characteristic of the Food Guide?
    a. It is used primarily by diabetics.
    b. It defines food portions according to energy content.
    c. The groupings are made according to nutrient content.
    d. It subdivides the grain group based on fiber content.

15. What is the recommendation regarding grain products?
    a. Consume 3 whole-grain products per day.
    b. ½ of total grains should be whole grains.

    c. Eat 6 enriched products a day.
    d. Take 2 tablespoons of laxatives each day.

16. Which of the following is a characteristic of enriched grain products?
    a. They have all of the added nutrients listed on the label.
    b. They have the fiber restored from the refining procedure.
    c. They have 4 vitamins and 4 minerals added by the food processor.
    d. They have virtually all the nutrients restored from the refining procedure.

17. When shopping for grains, this type of product may be rich in all nutrients found in original grain:
    a. refined.   c. whole grain.
    b. enriched.   d. fiber rich.

18. Dried beans and peas, pinto beans, lima beans and black beans are examples of:
    a. legumes.   c. milk alternatives.
    b. meats.   d. fruits.

19. Whole-grain bread contains more of the following nutrients than enriched bread:
    a. iron, carbohydrate, and protein.   c. magnesium, zinc, fiber and vitamin B_6.
    b. iron, thiamin, riboflavin, and niacin.   d. water, fiber, and fat.

20. Corn, green peas, and potatoes are listed with breads in which food plan?
    a. Food Guide   c. Healthy Eating Index
    b. Exchange lists   d. Four food group plan

21. Foods that substitute for and resemble another food but are nutritionally inferior to it are called:
    a. functional foods.   c. food density.
    b. food substitutes.   d. imitation foods.

22. Another name for fat-free milk is:
    a. nonfat milk.   c. reduced-fat milk.
    b. low-fat milk.   d. less-fat milk.

23. Wise consumers will take this action before using soy products as substitutes for dairy:
    a. check expiration dates.
    b. use soy milk even if they dislike the taste.
    c. check to see if they are organic.
    d. read labels to see if they have been fortified with vitamins B_{12} and D and calcium.

24. The items on a food label that tell consumers about the nutritional value of a product include:
    1. the common or usual name of the product.
    2. the net contents in terms of weight, measure or count.
    3. the ingredients in descending order of predominance.
    4. the serving size and number of servings per container.
    5. the quantities of specified nutrients and food constituents.

    a. 1, 2, 3, and 4   c. 3, 4, and 5
    b. 2, 3, 4, and 5   d. all of the above

25. Serving sizes on a label:
    a. vary according to brand.
    b. are not always the same as those of the USDA Food Guide.
    c. are 1 cup for all ice creams.
    d. b and c

26. The FDA uses _____ kcalories as a standard for energy intake in calculating the Daily Values (DV) for energy-yielding nutrients.
    a. 1,200   c. 1,800
    b. 1,500   d. 2,000

27. If you see "good source of fiber" on a label, this is an example of:
    a. a nutrient claim.   c. a structure-function claim.
    b. a health claim.   d. all of the above

28. If you see "Diets low in sodium may reduce the risk of high blood pressure," on a label, this is an example of:
    a. a nutrient claim.   c. a structure-function claim.
    b. a health claim.   d. all of the above

29. If you read, "builds strong bones," on a label, this is an example of:
    a. a nutrient claim.   c. a structure-function claim.
    b. a health claim.   d. all of the above

30. If a food product provides between 10 and 19 percent of the Daily Value for a nutrient, this is considered a(n) _____ source.
    a. high   c. good
    b. excellent   d. low

Short Answer Questions

1. Diet-planning principles include:

    a.   d.

    b.   e.

    c.   f.

2. The two diet-planning guides most widely used are:

    a.

    b.

3. The five groups in the USDA Food Guide are:

    a.   d.

    b.   e.

    c.

4. The five vegetable subgroups are:

    a.   d.

    b.   e.

    c.

Problem Solving

1. Using the exchange system, how many kcalories are in the following breakfast?

    1 slice of whole-wheat toast with
    1 tsp butter
    1 small banana, ½ cup apple juice
    1 ½ cups nonfat milk

2. A food item provides 400 mg of calcium and has 350 kcalories. Calculate the nutrient density value.

3. One food item provides 10 mg of iron and 100 kcalories. Another food item provides 15 mg of iron and 175 kcalories. Calculate the nutrient density for both food items. Which is a better choice?

4. A recipe calls for 2 teaspoons of sugar. Calculate this amount in milliliters (mL).

5. You are trying to consume 8 cups of water and day. How many milliliters (mL) is this?

6. Suppose your energy requirements are 2500 kcal per day. Calculate your personal Daily Value for g of fat per day.

7. Suppose your energy requirements are 1800 kcal per day. Calculate your personal Daily Value for g of fat per day.

8. Your friend's energy requirements are 3000 kcal per day. How many grams of saturated fat, carbohydrate, fiber, and protein would he require, based on personalized Daily Values?

9. The serving size for a food item is ¾ cup. You have consumed 3 cups. How many servings did you eat?

10. The serving size for a food item is ¾ cup and each serving has 120 kcalories. You have consumed 3 cups. How many kcalories did the food provide?

Crossword Puzzle

Complete this crossword puzzle by Mary A. Wyandt, Ph.D., CHES.

Across	Down
1. statements that characterize the quantity of a nutrient in a food.	2. a tan-colored endosperm flour with texture and nutritive qualities that approximate those of regular white flour
3. an elastic protein found in wheat and other grains that gives dough its structure and cohesiveness	4. plants of the bean and pea family, rich in high-quality protein compared with other plant-derived foods
7. the outer, inedible part of a grain; also called the chaff	5. a grain milled in its entirety (all but the husk); not refined
8. the bulk of the edible part of the kernel containing starch and protein	6. the protective coating around the kernel similar in function to the shell of a nut, rich in nutrients and fiber
10. a measure of the nutrients a food provides relative to the energy it provides	9. the nutrient-rich inner part of a grain; the seed that grows into a wheat plant

⑥ Chapter 2 Answer Key ⑥

Summing Up

1. adequacy	9. nutrient	17. amounts	25. milk
2. balance	10. *Guidelines*	18. protein	26. diets
3. kcalorie	11. balanced	19. carbohydrate	27. descending
4. variety	12. five	20. fat	28. Serving
5. variety	13. nutrient	21. sizes	29. percent
6. nutrients	14. lacking	22. kcalorie	30. educate.
7. weight;	15. nutrient-dense	23. whole-grain	
8. exchange	16. discretionary	24. vegetables	

Chapter Study Questions

1. Adequacy—providing all essential nutrients. Balance—providing foods of a number of types in proportion to each other so that foods rich in some nutrients do not crowd out foods rich in other nutrients. Kcalorie control—management of food energy intake. Nutrient density—selecting foods with high nutrient value relative to food energy. Moderation—providing enough but not too much of a dietary constituent. Variety—using different foods on different occasions; variety helps ensure adequacy and balance.

2. See Table 2-1.

3. See Figure 2-1 for food groups and foods typical of each group. Groups foods similar in origin and that make notable contributions of the same key nutrients. Nutrient density and kcalorie control are addressed in that the plan points out foods that are more and less nutrient dense; the plan also gives specific recommended amounts for people with different kcalorie needs. Additionally, this plan specifies the amounts that can be allotted for a discretionary kcalorie allowance.

4. See Figure 2-1 (answers are individual).

5. Notable nutrients: nutrients that are most often lacking in the diets of Americans, including fiber, vitamin A, vitamin C, vitamin E, calcium, magnesium, and potassium Nutrient density: a measure of the nutrients a food provides relative to the energy it provides; the more nutrients and fewer kcalories, the higher the nutrient density. Discretionary kcalorie allowance: the kcalories remaining in a person's energy allowance after consuming enough nutrient-dense foods to meet all nutrient needs for a day.

6. MyPyramid is a diet planning program that can be used to plan a healthy diet. It educates consumers about healthy food and physical activity choices, and it helps consumers choose the kinds and amounts of foods to eat each day.

7. Exchange lists are used for kcalorie control. They group foods according to kcalorie content, and consumers can select foods from exchange lists that are significant to energy content.

8. The common or usual name of the product; the name and address of the manufacturer, packer, or distributor; the net contents in terms of weight, measure, or count; the ingredients in descending order of predominance by weight; the serving size and number of servings per container; and the quantities of specified nutrients and food constituents. Consumers can use ingredient lists to compare the nutrient density of products.

9. Daily Values are reference values developed by the FDA for use on food labels; they help people compare foods to their recommended intakes.

10. Nutrient claims: statements that characterize the quantity of a nutrient in a food. Health claims: statements that characterize the relationship between a nutrient or other substance in a food and a disease or health-related condition. Structure-function claims: statements that characterize the relationship between a nutrient or other substance in a food and its role in the body.

Sample Test Questions

1. c (p. 37)	9. a (p. 47)	17. c (p. 51)	25. b (p. 55)
2. e (p. 38)	10. d (p. 44)	18. a (pp. 44, 52)	26. d (p. 56)
3. b (p. 38)	11. d (p. 44)	19. c (p. 51)	27. a (p. 58)
4. a (p. 39)	12. c (p. 45)	20. b (p. 48)	28. b (p. 59)
5. d (p. 39)	13. c (p. 47)	21. d (p. 53)	29. c (p. 59)
6. e (p. 44)	14. c (pp. 41, 44)	22. a (p. 53)	30. c (pp. 57, 58)
7. c (p. 41)	15. b (p. 40)	23. d (pp. 65-67)	
8. a (p. 41)	16. a (pp. 50-51)	24. c (pp. 55-56)	

Short Answer Questions

1. adequacy, balance, kcalorie control, nutrient density, moderation, variety
2. food group plans; exchange systems
3. fruits, vegetables, grains, meat and legumes, milk
4. dark green, orange and deep yellow, legumes, starchy vegetables, other vegetables

Problem Solving

1. $80 + 45 + 2(60) + 1.5(90) = 380$ kcal
2. 400 mg calcium divided by 350 kcal = 1.14 mg calcium per kcalorie
3. 10 mg iron divided by 100 kcal = 0.1; 15 mg iron divided by 175 kcalories = .085.
   The first item has greater nutrient density, and based on this information is a better choice.
4. 2 teaspoons x 5 mL = 10 milliliters (mL)
5. 8 cups x 240 mL in a cup = 1920 milliliters (mL)
6. 2500 x 0.30 kcal from fat = 750 kcal divided by 9 kcal/g = 83 grams of fat
7. 1800 x 0.30 kcal from fat = 540 kcal divided by 9 kcal/g = 60 grams of fat
8. 3000 x 0.10 = 300 kcal from saturated fat; 300 kcal divided by 9 = 33 grams saturated fat per day
   3000 x 0.60 = 1800 kcal from carbohydrate; 1800 divided by 4 = 450 grams carbohydrate
   3000 divided by 1000 kcal = 3.0; 3.0 x 11.5 g = 34.5 grams fiber
   3000 x 0.10 = 300 kcal divided by 4 = 75 grams of protein
9. 3 cups divided by .75 cup per serving = 4 servings
10. 3 cups divided by .75 cup per serving = 4 servings x 120 kcalories = 480 kcalories

Crossword Puzzle

1. nutrient claims	4. legumes	7. husk	10. nutrient density
2. unbleached flour	5. whole grain	8. endosperm	
3. gluten	6. bran	9. germ	

⊚ Chapter 3 – Digestion, Absorption, and Transport ⊚

Chapter Outline

I. Digestion
  A. Anatomy of the Digestive Tract
    1. Mouth
    2. Esophagus to the Stomach
    3. Small Intestine
    4. Large Intestine (Colon)
  B. The Muscular Action of Digestion
    1. Peristalsis
    2. Stomach Action
    3. Segmentation
    4. Sphincter Contractions
  C. The Secretions of Digestion
    1. Saliva
    2. Gastric Juice
    3. Pancreatic Juice and Intestinal Enzymes
    4. Bile
  D. The Final Stage
II. Absorption
  A. Anatomy of the Absorptive System
  B. A Closer Look at the Intestinal Cells

    1. Specialization in the GI Tract
    2. The Myth of "Food Combining"
    3. Preparing Nutrients for Transport
III. The Circulatory Systems
  A. The Vascular System
  B. The Lymphatic System
IV. The Health and Regulation of the GI Tract
  A. Gastrointestinal Bacteria
  B. Gastrointestinal Hormones and Nerve Pathways
  C. The System at Its Best
V. Common Digestive Problems
  A. Choking
  B. Vomiting
  C. Diarrhea
  D. Constipation
  E. Belching and Gas
  F. Heartburn and "Acid Indigestion"
  G. Ulcers

Summing Up

The gastrointestinal tract is a flexible, muscular tube that facilitates the processes of digestion and (1)_____. Substances that penetrate the tract's wall will enter the body, but many things pass through unabsorbed. Food enters the mouth past the epiglottis, then travels to the (2)_____, through the cardiac sphincter to the stomach, through the (3)_____ to the small intestine (the duodenum, with entrance from the gallbladder and (4)_____; then the jejunum; then the ileum), through the ileocecal valve to the large intestine, and finally past the (5)_____ to the rectum, ending at the anus.

Involuntary muscles and (6)_____ of the digestive tract function without any conscious effort. Peristalsis and (7)_____ mix and move intestinal contents along the gastrointestinal tract. Digestive (8)_____ cause food substances to break down into simpler compounds. Nutrients are then (9)_____ in the microvilli of the (10)_____ intestine. The nutrients are transported to various parts of the body by the circulatory systems—the (11)_____ system and the lymphatic system. The (12)_____ and nervous systems coordinate the digestive and absorptive processes. Characteristics of meals that promote optimal absorption of nutrients are balance, (13)_____, adequacy, and (14)_____.

Chapter Study Questions

1. Describe the obstacles associated with digesting food and the solutions offered by the human body.

2. Describe the path food follows as it travels through the digestive system. Summarize the muscular actions that take place along the way.

3. Name five organs that secrete digestive juices. How do the juices and enzymes facilitate digestion?

4. Describe the problems involved with absorbing nutrients and the solutions offered by the small intestine.

5. How is blood routed through the digestive system? Which nutrients enter the bloodstream directly? Which are first absorbed into the lymph?

6. Describe how the body coordinates and regulates the processes of digestion and absorption.

7. How does the composition of the diet influence the functioning of the GI tract?

8. What steps can you take to help your GI tract function at its best?

Chapter Glossary

- **anus:** the terminal outlet of the GI tract.
- **appendix:** a narrow blind sac extending from the beginning of the colon that stores lymph cells.
- **arteries:** vessels that carry blood from the heart to the tissues.
- **bolus:** a portion; with respect to food, the amount swallowed at one time.
- **-ase:** a word ending denoting an enzyme. The word beginning often identifies the compounds the enzyme works on.
- **bicarbonate:** an alkaline compound with the formula HCO_3 that is secreted from the pancreas as part of the pancreatic juice. (Bicarbonate is also produced in all cell fluids from the dissociation of carbonic acid to help maintain the body's acid-base balance.)
- **bile:** an emulsifier that prepares fats and oils for digestion; an exocrine secretion made by the liver, stored in the gallbladder, and released into the small intestine when needed.
- **capillaries:** small vessels that branch from an artery. Capillaries connect arteries to veins. Exchange of oxygen, nutrients, and waste materials takes place across capillary walls.
- **carbohydrase:** an enzyme that hydrolyzes carbohydrates.
- **cardiac sphincter:** the sphincter muscle at the junction between the esophagus and the stomach (also called the *lower esophageal sphincter*).
- **catalyst:** a compound that facilitates chemical reactions without itself being changed in the process.
- **cholecystokinin (CCK):** a hormone produced by cells of the intestinal wall. Target organ: the gallbladder. Response: release of bile and slowing of GI motility.
- **chyme:** the semiliquid mass of partly digested food expelled by the stomach into the duodenum.
- **colon:** the lower portion of the intestine that completes the digestive process; its segments are the ascending colon, the transverse colon, the descending colon, and the sigmoid colon.
- **crypts:** tubular glands that lie between the intestinal villi and secrete intestinal juices into the small intestine.
- **digestion:** the process by which food is broken down into absorbable units.
- **digestive enzymes:** proteins found in digestive juices that act on food substances, causing them to break down into simpler compounds.
- **digestive system:** all the organs and glands associated with the ingestion and digestion of food.
- **duodenum:** the top portion of the small intestine.
- **emulsifier:** a substance with both water-soluble and fat-soluble portions that promotes the mixing of oils and fats in a watery solution.
- **endocrine glands:** glands that secrete their materials into the blood.
- **enterogastrones:** gastrointestinal hormones; specifically, any hormone that slows motility and inhibits gastric secretions.
- **epiglottis:** cartilage in the throat that guards the entrance to the trachea and prevents fluid or food from entering it when a person swallows.
- **esophageal sphincter:** a sphincter muscle at the upper or lower end of the esophagus.
- **esophagus:** the food pipe; the conduit from the mouth to the stomach.
- **exocrine glands:** glands that secrete their materials into the digestive tract or onto the surface of the skin.
- **feces:** waste matter discharged from the colon; also called *stools*.
- **flora:** bacteria in the intestines; also called *microflora*.
- **gallbladder:** the organ that stores and concentrates bile. When it receives the signal that fat is present in the duodenum, the gallbladder contracts and squirts bile through the bile duct into the duodenum.
- **gastric glands:** exocrine glands in the stomach wall that secrete gastric juice into the stomach.
- **gastric juice:** the digestive secretion of the gastric glands of the stomach.
- **gastrin:** a hormone secreted by cells in the stomach wall. Target organ: the glands of the stomach. Response: secretion of gastric acid.
- **gastrointestinal (GI) tract:** the digestive tract. The principal organs are the stomach and intestines.
- **gland:** a cell or group of cells that secretes materials for special uses in the body.
- **goblet cells:** cells of the GI tract (and lungs) that secrete mucus.
- **hepatic portal vein:** the vein that collects blood from the GI tract and conducts it to capillaries in the liver.

- **hepatic vein:** the vein that collects blood from the liver capillaries and returns it to the heart.
- **homeostasis:** the maintenance of constant internal conditions (such as blood chemistry, temperature, and blood pressure) by the body's control systems. A homeostatic system is constantly reacting to external forces to maintain limits set by the body's needs.
- **hormones:** chemical messengers. Hormones are secreted by a variety of glands in response to altered conditions in the body. Each hormone travels to one or more specific target tissues or organs, where it elicits a specific response to maintain homeostasis.
- **hydrochloric acid:** an acid composed of hydrogen and chloride atoms (HCl), normally produced by the gastric glands.
- **hydrolysis:** a chemical reaction in which a major reactant is split into two products, with the addition of a hydrogen atom (H) to one and a hydroxyl group (OH) to the other (from water, H_2O). (The noun is *hydrolysis*; the verb is *hydrolyze*.)
- **ileocecal valve:** the sphincter separating the small and large intestines.
- **ileum:** the last segment of the small intestine.
- **jejunum:** the first two-fifths of the small intestine beyond the duodenum.
- **large intestine:** the lower portion of intestine that completes the digestive process; its segments are the ascending colon, the transverse colon, the descending colon, and the sigmoid colon; also called *colon*.
- **lipase:** an enzyme that hydrolyzes lipids (fats).
- **liver:** the organ that manufacturers bile.
- **lumen:** the space within a vessel, such as the intestine.
- **lymph:** a clear yellowish fluid that is similar to blood except that it contains no red blood cells or platelets. Lymph from the GI tract transports fat and fat-soluble vitamins to the bloodstream via lymphatic vessels.
- **lymphatic system:** a loosely organized system of vessels and ducts that convey fluids toward the heart. The GI part of the lymphatic system carries the products of fat digestion into the bloodstream.
- **microvilli:** tiny, hairlike projections on each cell of every villus that can trap nutrient particles and transport them into the cells; singular *microvillus*.
- **motility:** the ability of the GI tract muscles to move.
- **mucus:** a slippery substance secreted cells of the GI lining (and other body linings) that protects the cells from exposure to digestive juices (and other destructive agents). The lining of the GI tract with its coat of mucus is a *mucous membrane*. (The noun is *mucus*; the adjective is *mucous*.)
- **pancreas:** a gland that secretes digestive enzymes and juices into the duodenum. The pancreas also secretes hormones that help to maintain glucose homeostasis.
- **pancreatic juice:** the exocrine secretion of the pancreas, containing enzymes for the digestion of carbohydrate, fat, and protein as well as bicarbonate, a neutralizing agent. The juice flows from the pancreas into the small intestine through the pancreatic duct. (The pancreas also has an endocrine function, the secretion of insulin and other hormones.)
- **peristalsis:** wavelike muscular contractions of the GI tract that push its contents along.
- **pH:** the unit of measure expressing a substance's acidity or alkalinity.
- **pharynx:** the passageway leading from the nose and mouth to the larynx and esophagus, respectively.
- **probiotics:** living microorganisms found in foods that, when consumed in sufficient quantities, are beneficial to health.
- **protease:** an enzyme that hydrolyzes proteins.
- **pyloric sphincter:** the circular muscle that separates the stomach from the small intestine and regulates the flow of partially digested food into the small intestine; also called *pylorus* or *pyloric valve*.
- **rectum:** the muscular terminal part of the intestine, extending from the sigmoid colon to the anus.
- **reflux:** a backward flow.
- **saliva:** the secretion of the salivary glands. Its principal enzyme begins carbohydrate digestion.
- **salivary glands:** exocrine glands that secret saliva into the mouth.
- **secretin:** a hormone produced by cells in the duodenum wall. Target organ: the pancreas. Response: secretion of bicarbonate-rich pancreatic juice.
- **segmentation:** a periodic squeezing or partitioning of the intestine at intervals along its length by its circular muscles.

- **small intestine:** a 10-foot length of small-diameter intestine that is the major site of digestion of food and absorption of nutrients; its segments are the duodenum, jejunum, and ileum.
- **sphincter:** a circular muscle surrounding, and able to close, a body opening. Sphincters are found at specific points along the GI tract and regulate the flow of food particles.
- **stomach:** a muscular, elastic, saclike portion of the digestive tract that grinds and churns swallowed food, mixing it with acid and enzymes to form chyme.
- **stools:** waste matter discharged from the colon; also called *feces*.
- **subclavian vein:** the vein that provides passageway from the lymphatic system to the vascular system.
- **thoracic duct:** the main lymphatic vessel that collects lymph and drains into the left subclavian vein.
- **trachea:** the air passageway from the larynx to the lungs; also called the *windpipe*.
- **veins:** vessels that carry blood to the heart.
- **villi:** fingerlike projections from the folds of the small intestine; singular *villus*.
- **yogurt:** milk product that results from the fermentation of lactic acid in milk by *Lactobacillus bulgaricus* and *Streptococcus thermophilus*.

Sample Test Questions

Select the best answer for each question.

1. The process by which food is broken down into absorbable units is called:
   a. absorption.
   b. digestion.
   c. peristalsis.
   d. reflux.

2. The uptake of nutrients by the cells in the small intestine for transport is called:
   a. absorption.
   b. digestion.
   c. bolus.
   d. segmentation.

3. The flexible muscular tube that extends from the mouth through the esophagus, stomach, small intestine, large intestine, and rectum to the anus is called:
   a. the lumen.
   b. the gastrointestinal tract.
   c. the bolus.
   d. segmentation.

4. The process of digestion begins in the:
   a. mouth.
   b. stomach.
   c. esophagus.
   d. small intestine.

5. The process of chewing is called:
   a. savory.
   b. mastication.
   c. eating.
   d. the bolus.

6. After a mouthful of food has been swallowed, it is called:
   a. a crypt.
   b. chyme.
   c. a bolus.
   d. stool.

7. To keep food from entering the lungs, the _____ closes off air passages.
   a. pharynx
   b. diaphragm
   c. esophageal sphincter
   d. epiglottis

8. The wavelike muscular contractions that propel food through the digestive tract are called:
   a. defecation.
   b. peristalsis.
   c. hydrolysis.
   d. denaturation.

9. Which sphincter muscle is situated between the stomach and the small intestine?
    a. cardiac sphincter
    b. pyloric sphincter
    c. ileocecal valve
    d. rectal sphincter

10. The partially digested food that enters the small intestine from the stomach is called:
    a. micelle.
    b. bile.
    c. chyme.
    d. feces.

11. An enzyme that hydrolyzes proteins is called:
    a. bile.
    b. a sphincter valve.
    c. pancreatic amylase.
    d. a protease.
    e. a secretory peptide.

12. The main function of bile is to:
    a. emulsify fats.
    b. stimulate the activity of protein digestive enzymes.
    c. neutralize the contents of the intestine.
    d. increase peristalsis.
    e. a, b, and c

13. Nutrients that are digested in the small intestine are:
    a. carbohydrate, fat, and protein.
    b. fat, water, and fiber.
    c. protein, vitamins, and fiber.
    d. water, fiber, and minerals.

14. A narrow blind sac extending from the beginning of the colon is called the:
    a. epiglottis.
    b. anus.
    c. rectum.
    d. appendix.

15. A single villus is composed of hundreds of _____, each covered with _____.
    a. organisms; mucus
    b. villi; lymph
    c. enzymes; mucus
    d. cells; microvilli
    e. crypts; glands

16. A group of cells that secretes materials for special uses in the body is:
    a. a gland.
    b. a sphincter.
    c. an enzyme.
    d. bile.

17. Avoiding certain food combinations at the same meal is:
    a. wise because the body cannot handle more than one task at a time.
    b. foolish because the body is equipped to handle digestion of a variety of foods and food types.
    c. practical and easy to adhere to.
    d. a and c

18. A periodic squeezing of the intestines is:
    a. peristalsis.
    b. diverticulosis.
    c. segmentation.
    d. defecation.

19. The small intestine is about _____ feet long.
    a. 10
    b. 30
    c. 50
    d. 100

20. The ability of the GI tract to move is called:
    a. peristalsis.
    b. reflux.
    c. motility.
    d. segmentation.

21. Glands that secrete intestinal juices into the small intestine are called:
    a. reflux.
    b. microvilli.
    c. villi.
    d. crypts.

22. The strength of acids is measured in:
    a. ion concentrations.
    b. alkaline tests.
    c. base assessments.
    d. pH units.

23. Living microorganisms found in foods that benefit health are called:
    a. yogurt.
    b. probiotics.
    c. antibiotics.
    d. prebiotics.

24. Waste matter discharged from the colon is:
    a. a bolus.
    b. chyme.
    c. lymph.
    d. stools.

25. Saliva contains:
    a. protease, bicarbonate, and water.
    b. bile, HCl, and water.
    c. chyme, pepsin, and bicarbonate.
    d. salts, carbohydrases, and water.
    e. a and b

26. The lymphatic system:
    a. contains red blood cells.
    b. eventually drains into the blood circulatory system.
    c. collects in a large duct behind the heart.
    d. a and b
    e. b and c

27. Which type of cells produce mucus that protects the stomach walls from gastric juices?
    a. secretin
    b. bile
    c. cholecystokinin
    d. goblet
    e. enterogastrone

28. What hormone is released in the presence of fat and slows intestinal motility to allow a longer digestion time?
    a. secretin
    b. bile
    c. cholecystokinin
    d. gastrin
    e. enterogastrone

29. The greatest danger from prolonged vomiting is:
    a. exhaustion.
    b. excess loss of fluid and salts.
    c. vitamin deficiencies.
    d. starvation.

30. A sensible idea for preventing constipation is to:
    a. take medication on a regular basis.
    b. cut down on water intake.
    c. include more high-fiber foods in the diet.
    d. include fewer high-fiber foods in the diet.

Short Answer Questions

1. Some problems involved in digestion include:

    a.

    b.

    c.

d.

e.

f.

g.

Figure Identification/Table Completion

A.   Identify the parts of the GI tract.

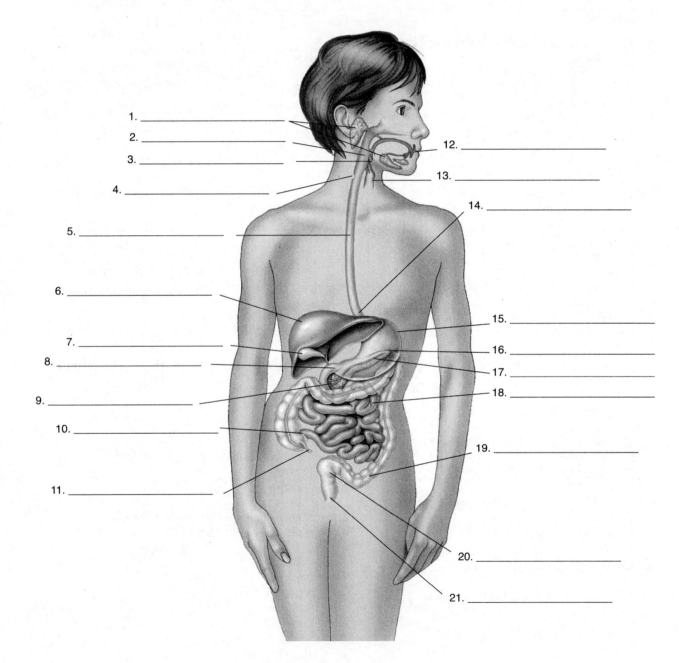

1. _____

2. _____

3. _____

4. _____

5. _____

6. _____

7. _____

8. _____

9. _____

10. _____

11. _____

12. _____

13. _____

14. _____

15. _____

16. _____

17. _____

18. _____

19. _____

20. _____

21. _____

B. Complete this chart identifying digestive secretions and their primary actions.

Organ or Gland	Target Organ	Secretion	Action
Salivary glands	Mouth	1._____	Fluid eases swallowing; salivary enzyme breaks down some 2._____.
Gastric glands	3._____	Gastric juice	Fluid mixes with bolus; hydrochloric acid uncoils 4._____; enzymes break down 5._____; 6._____ protects stomach cells.
Pancreas	Small intestine	7._____	8._____ neutralizes acidic gastric juices; pancreatic 9._____ break down carbohydrates, fats, and proteins.
Liver	10._____	Bile	Bile stored until needed.
Gallbladder	11._____	Bile	Bile 12._____ fat so that enzymes can attack.
Intestinal glands	Small intestine	Intestinal juice	Intestinal 13._____ break down carbohydrate, fat, and protein fragments; 14._____ protects the intestinal wall.

35

C. Complete the chart summarizing the chemical digestion and absorption that occur in each part of the GI tract.

	Mouth	Stomach	Small Intestine	Large Intestine
Carbohydrate				
Fiber				
Protein				
Fat				
Vitamins				
Minerals				

Crossword Puzzle

Complete this crossword puzzle by Mary A. Wyandt, Ph.D., CHES.

Across	Down
1. an enzyme that hydrolyzes proteins	2. passage of nutrients from the GI tract into either the blood or lymph
4. a substance with both water-soluble and fat-soluble portions that promotes the mixing of oils and fats in a watery solution	3. a chemical reaction in which a major reactant is split into two products, with the addition of H to one and OH to the other (from water)
6. a hormone secreted by cells in the stomach wall	5. an enzyme that hydrolyzes lipids (fats)
7. a portion; with respect to food, amount swallowed at one time	7. an emulsifier that prepares fats and oils for digestion; an exocrine secretion made by the liver and stored in the gallbladder
8. a hormone produced by cells in the duodenum wall	9. a backward flow

⑥ Chapter 3 Answer Key ⑥

Summing Up

1. absorption
2. esophagus
3. pylorus
4. pancreas

5. appendix
6. glands
7. segmentation
8. enzymes

9. absorbed
10. small
11. vascular
12. endocrine

13. variety
14. moderation

Chapter Study Questions

1. The epiglottis closes off the airways so that food and liquid do not enter the lungs. Food passes through the diaphragm to reach the stomach by way of the esophagus. Fluids are added to the food as it travels through the system allowing smooth passage. Water is reabsorbed in the large colon, thus conserving water and creating a semisolid waste. Peristalsis keeps the materials steadily moving through the system and sphincter muscles serve as one-way gates allowing small quantities to pass at appropriate intervals. Stomach cells secrete mucus to protect them from acid and enzymes that would digest them. Rectal muscles prevent elimination until voluntarily performed.

2. Food enters the mouth and travels past the epiglottis, down the esophagus and through the cardiac sphincter to the stomach, then through the pyloric sphincter to the small intestine, on through the ileocecal valve to the large intestine, past the appendix to the rectum, ending at the anus. Muscular actions include: chewing, swallowing, peristalsis, segmentation, and sphincter contractions.

3. Salivary glands, stomach glands, pancreas, liver, gallbladder, and intestinal glands. Salivary glands secrete saliva that contains amylase enzyme that breaks down starch. Gastric juice is secreted by the cells in the stomach wall and contains pepsin and HCl that break down proteins. Pancreatic juice contains bicarbonate that neutralizes acidic gastric juices as well as other enzymes that break down CHO, protein and fat. The gallbladder secretes bile which emulsifies fat.

4. The body must find a way to absorb many molecules. It solves this by its anatomy—the small intestine has hundreds of folds, each covered with thousands of villi, which in turn are composed of hundreds of cells, which in turn are covered with microvilli, providing a very large surface area for nutrient molecules to make contact and be absorbed.

5. Heart to arteries to capillaries (in intestines) to vein to capillaries (in liver) to vein to heart. Water-soluble nutrients and small products of fat digestion enter the bloodstream directly; large fats and fat-soluble nutrients are first absorbed into the lymph.

6. The body's hormonal system and nervous system coordinate all the digestive and absorptive processes. The contents in the GI tract either stimulate or inhibit digestive secretions by way of messages that are carried from one section of the GI tract to another by both hormones and nerve pathways.

7. Enzyme activity changes proportionately in response to the amounts of carbohydrate, fat and protein in the diet. Hormones in the GI tract inform the pancreas as to the amount and type of enzymes to secrete in response to diet; the presence of fat slows GI motility.

8. Obtain adequate sleep, engage in physical activity, keep a positive state of mind, and eat meals with these characteristics: balance, moderation, variety, adequacy.

Sample Test Questions

1. b (p. 71)
2. a (p. 71)
3. b (p. 72)
4. a (p. 72)
5. b (p. 72)
6. c (p. 73)
7. d (p. 73)
8. b (p. 74)

9. b (pp. 72, 74)
10. c (p. 74)
11. d (p. 77)
12. a (p. 78)
13. a (pp. 77, 79)
14. d (p. 74)
15. d (p. 80)
16. a (p. 78)

17. b (pp. 81-82)
18. c (pp. 74, 75-76)
19. a (p. 80)
20. c (p. 74)
21. d (p. 80)
22. d (p. 77)
23. b (p. 86)
24. d (p. 78)

25. d (p. 77)
26. e (p. 85)
27. d (pp. 77, 80)
28. c (p. 88)
29. b (p. 93)
30. c (p. 95)

Short Answer Questions

1. a. Air and food must both go to the stomach but food and liquid cannot go to the lungs.
   b. Food must be conducted through the diaphragm to reach the abdomen.
   c. The amount of water should be regulated to keep the intestinal contents at the right consistency.
   d. Water must be withdrawn from the intestinal contents after absorption.
   e. The materials within the tract should move steadily except when a poison has been swallowed (in this case the contents should reverse direction and move quickly). If infection sets in farther down the tract, the flow should be accelerated.
   f. The cells of the digestive tract need protection against the powerful juices they secrete.
   g. Provision must be made for periodic, voluntary evacuation when convenient.

Figure Identification/Table Completion

A.
1. salivary glands
2. pharynx
3. epiglottis
4. upper esophageal sphincter
5. esophagus
6. liver
7. gallbladder
8. pyloric sphincter
9. bile duct
10. ileocecal valve
11. appendix
12. mouth
13. trachea
14. lower esophageal sphincter
15. stomach
16. pancreas
17. pancreatic duct
18. small intestine
19. large intestine
20. rectum
21. anus

B.
1. Saliva
2. carbohydrate
3. Stomach
4. proteins
5. proteins
6. mucus
7. Pancreatic juice
8. Bicarbonate
9. enzymes
10. Gallbladder
11. Small intestine
12. emulsifies
13. enzymes
14. mucus

C.

Mouth	Stomach	Small Intestine	Large Intestine
Carbohydrate Salivary enzyme begins to break down starch.	Digestion continues until HCl inactivates the salivary enzyme, then stops.	Sugars are absorbed. Pancreatic enzymes resume starch digestion. Intestinal cell enzymes complete starch digestion, and the resulting small fragments are absorbed into the portal vein.	None.
Fiber None.	None.	None.	Some fibers are partially digested by bacteria; some of the products are absorbed. Most fibers are simply excreted.
Protein None.	Begin to uncoil when mixed with gastric acid. Then gastric protease enzymes begin digestion.	Pancreatic and intestinal proteases complete digestion, and resulting small fragments are absorbed into the portal vein.	None.
Fat Minimal in adults.	None.	Bile emulsifies fat. Pancreatic and intestinal lipases then digest it into small fragments that are absorbed into the lymph.	No digestion or absorption occur. Some fat and cholesterol bind to fiber and are excreted.
Vitamins None.	None	Vitamins are absorbed.	None.
Minerals None.	None	Minerals are absorbed.	Some minerals are absorbed; others bind to fiber and are excreted.

Crossword Puzzle

1. protease
2. absorption
3. hydrolysis
4. emulsifier
5. lipase
6. gastrin
7D. bile
7A. bolus
8. secretin
9. reflux

⑥ Chapter 4 – The Carbohydrates: ⑥ Sugars, Starches, and Fibers

Chapter Outline

I.   The Chemist's View of Carbohydrates
II. Simple Carbohydrates
    A. Monosaccharides
       1. Glucose
       2. Fructose
       3. Galactose
    B. Disaccharides
       1. Condensation
       2. Hydrolysis
       3. Maltose
       4. Sucrose
       5. Lactose
III. The Complex Carbohydrates
    A. Glycogen
    B. Starches
    C. Fibers
       1. Soluble Fibers
       2. Insoluble Fibers
       3. Fiber Sources
       4. Resistant Starches
       5. Phytic Acid
IV. Digestion and Absorption of Carbohydrates
    A. Carbohydrate Digestion
       1. In the Mouth
       2. In the Stomach
       3. In the Small Intestine
       4. In the Large Intestine
    B. Carbohydrate Absorption
    C. Lactose Intolerance
       1. Symptoms
       2. Causes
       3. Prevalence
       4. Dietary Changes
V.   Glucose in the Body
    A. A Preview of Carbohydrate Metabolism
       1. Storing Glucose as Glycogen
       2. Using Glucose for Energy
       3. Making Glucose from Protein
       4. Making Ketone Bodies from Fat Fragments
       5. Using Glucose to Make Fat
       6. The Constancy of Blood Glucose
       7. Maintaining Glucose Homeostasis
       8. The Regulating Hormones
       9. Balancing within the Normal Range
      10. Falling Outside the Normal Range
      11. Diabetes
      12. Hypoglycemia
      13. The Glycemic Response

VI. Health Effects and Recommended Intakes of Sugars
    A. Health Effects of Sugars
       1. Nutrient Deficiencies
       2. Dental Caries
    B. Controversies Surrounding Sugars
       1. Controversy: Does Sugar Cause Obesity?
       2. Controversy: Does Sugar Cause Heart Disease?
       3. Controversy: Does Sugar Cause Misbehavior in Children and Criminal Behavior in Adults?
       4. Controversy: Does Sugar Cause Cravings and Addictions?
    C. Recommended Intakes of Sugars
VII. Health Effects and Recommended Intakes of Starch and Fibers
    A. Health Effects of Starch and Fibers
       1. Heart Disease
       2. Diabetes
       3. GI Health
       4. Cancer
       5. Weight Management
       6. Harmful Effects of Excessive Fiber Intake
    B. Recommended Intakes of Sugars and Fibers
    C. From Guidelines to Groceries
       1. Grains
       2. Vegetables
       3. Fruits
       4. Milks and Milk Products
       5. Meats and Meat Alternates
       6. Read Food Labels
VIII. Alternatives to Sugar
    A. Artificial Sweeteners
       1. Saccharin
       2. Aspartame
       3. Acesulfame-K
       4. Sucralose
       5. Neotame
       6. Tagatose
       7. Alitame and Cyclamate
       8. Acceptable Daily Intake
       9. Artificial Sweeteners and Weight Control
    B. Stevia—An Herbal Alternative
    C. Sugar Replacers

Summing Up

At least half our food energy is derived from (1)_____, principally from (2)_____, but also from the simple sugars. Carbohydrates are classified as (3)_____ carbohydrates [the (4)_____] or simple carbohydrates [the (5)_____ and disaccharides]. Each of the three disaccharides [(6)_____, (7)_____, and maltose] contains a molecule of glucose paired with either (8)_____, galactose, or another glucose.

The polysaccharides starch and glycogen are composed of chains of (9)_____ units. (10)_____ is the storage form of glucose in the plant. Sources of starch in the diet include seeds, (11)_____, and starchy vegetables. (12)_____, or animal starch, is more complex than starch and is synthesized in the (13)_____ and muscle from excess glucose in the bloodstream. The fibers include the polysaccharides (14)_____, pectin, and hemicellulose, mucilages, and gums as well as the nonpolysaccharide lignin.

Nutrition status affects our well-being even at the level of the body's (15)_____. The body strives to maintain its blood glucose within a normal range for optimal health and functioning. The hormones (16)_____, glucagon, and epinephrine function to maintain glucose (17)_____ in the body.

It is recommended that people consume (18)_____ to (19)_____ percent of their total kcalories from carbohydrates, preferably from (20)_____ carbohydrates. The (21)_____ system provides a useful guide to estimating the carbohydrate content of a meal.

Plant fibers also include the polysaccharides cellulose, (22)_____, and hemicellulose. There are also carbohydrate-like sources of fiber, such as (23)_____ and gums. The DRI for fiber is (24)_____ grams per (25)_____ kcalories consumed.

Chapter Study Questions

1. Which carbohydrates are described as simple and which are complex?

2. Describe the structure of a monosaccharide and name the three monosaccharides important in nutrition. Name the three disaccharides commonly found in foods and their component monosaccharides. In what foods are these sugars found?

3. What happens in a condensation reaction? In a hydrolysis reaction?

4. Describe the structure of polysaccharides and name the ones important in nutrition. How are starch and glycogen similar and how do they differ? How do the fibers differ from the other polysaccharides?

5. Describe carbohydrate digestion and absorption. What role does fiber play in the process?

6. What are the possible fates of glucose in the body? What is the protein-sparing action of carbohydrate?

7. How does the body maintain blood glucose concentrations? What happens when blood glucose rises too high or falls too low?

8. What are the health effects of sugars? What are the dietary recommendations regarding concentrated sugar intakes?

9. What are the health effects of starches and fibers? What are the dietary recommendations regarding these complex carbohydrates?

10. What foods provide starches and fibers?

Chapter Glossary

- **acid-base balance:** the equilibrium in the body between acid and base concentrations.
- **added sugars:** sugars and syrups used as an ingredient in the processing and preparation of foods such as breads, cakes, beverages, jellies, and ice cream as well as sugars eaten separately or added to foods at the table.
- **amylase:** an enzyme that hydrolyzes amylose (a form of starch). Amylase is a *carbohydrase*, an enzyme that breaks down carbohydrates.
- **carbohydrates:** compounds composed of carbon, oxygen, and hydrogen arranged as monosaccharides or multiples of monosaccharides. Most, but not all carbohydrates have a ratio of one carbon molecule to one water molecule: $(CH_2O)_n$.
- **complex carbohydrates (starches and fibers):** polysaccharides composed of straight or branced chains of monosaccharides.
- **condensation:** a chemical reaction in which two reactants combine to yield a larger product.
- **dental caries:** decay of teeth.
- **dental plaque:** a gummy mass of bacteria that grows on teeth and can lead to dental caries and gum disease.
- **diabetes:** a chronic disorder of carbohydrate metabolism, usually resulting from insufficient or ineffective insulin.
- **dietary fibers:** in plant foods, the *nonstarch polysaccharides* that are not digested by human digestive enzymes, although some are digested by GI tract bacteria. Dietary fibers include cellulose, hemicelluloses, pectins, gums, and mucilages and the nonpolysaccharides lignins, cutins, and tannins.
- **disaccharides:** pairs of monosaccharides linked together.
- **epinephrine:** a hormone of the adrenal gland that modulates the stress response; formerly called **adrenaline**. When administered by injection, epinephrine counteracts anaphylactic shock by opening the airways and maintaining heartbeat and blood pressure.
- **fermentable:** the extent to which bacteria in the GI tract can break down fibers to fragments that the body can use.
- **fructose:** a monosaccharide. Sometimes known as frit sugar or **levulose**, fructose is found abundantly in fruits, honey, and saps.
- **galactose:** a monosaccharide; part of the disaccharide lactose.
- **glucagon:** a hormone that is secreted by special cells in the pancreas in response to low blood glucose concentration and elicits release of glucose from liver glycogen stores.
- **gluconeogenesis:** the making of glucose from a noncarbohydrate source.
- **glucose:** a monosaccharide; sometimes known as blood sugar or **dextrose**.
- **glycemic index:** a method of classifying foods according to their potential for raising blood glucose.
- **glycemic response:** the extent to which a food raises the blood glucose concentration and elicits an insulin response.
- **glycogen:** an animal polysaccharide composed of glucose; manufactured and stored in the liver and muscles as a storage form of glucose. Glycogen is not a significant food source of carbohydrate and is not counted as one of the complex carbohydrates in foods.
- **hypoglycemia:** an abnormally low blood glucose concentration.
- **insoluble fibers:** indigestible food components that do not dissolve in water. Examples include the tough, fibrous structures found in the strings of celery and the skins of corn kernels.
- **insulin:** a hormone secreted by special cells in the pancreas in response to (among other things) increased blood glucose concentration. The primary role of insulin is to control the transport of glucose from the bloodstream into the muscle and fat cells.
- **kefir:** a fermented milk created by adding *Lactobacillus acidophilus* and other bacteria that break down lactose to glucose and galactose, producing a sweet lactose-free product.
- **ketone bodies:** the product of the incomplete breakdown of fat when glucose is not available in the cells.
- **ketosis:** an undesirably high concentration of ketone bodies in the blood and urine.
- **lactase:** an enzyme that hydrolyzes lactose.
- **lactase deficiency:** a lack of enzyme required to digest the disaccharide lactose into its component monosaccharides (glucose and galactose).
- **lactose:** a disaccharide composed of glucose and galactose; commonly known as milk sugar.

44

- **lactose intolerance:** a condition that results from inability to digest the milk sugar lactose; characterized by bloating, gas, abdominal discomfort, and diarrhea. Lactose intolerance differs from milk allergy, which is caused by an immune reaction to the protein in milk.
- **maltase:** an enzyme that hydrolyzes maltose.
- **maltose:** a disaccharide composed of two glucose units; sometimes known as malt sugar.
- **monosaccharides:** carbohydrates of the general formula $C_nH_{2n}O_n$ that consist of a single ring.
- **oligosaccharide:** an intermediate string of three to ten monosaccharides.
- **phytic acid:** a nonnutrient component of plant seeds; also called **phytate**. Phytic acid occurs in the husks of grains, legumes, and seeds and is capable of binding minerals such as zinc, iron, calcium, magnesium, and copper in insoluble complexes in the intestine, which the body excretes unused.
- **polysaccharides:** compounds composed of many monosaccharides linked together.
- **protein-sparing action:** the action of carbohydrate (and fat) in providing energy that allows protein to be used for other purposes.
- **resistant starches:** starches that escape digestion and absorption in the small intestine of healthy people.
- **satiety:** the feeling of fullness and satisfaction that occurs after a meal and inhibits eating until the next meal. Satiety determines how much time passes between meals.
- **serotonin:** a neurotransmitter important in sleep regulation, appetite control, intestinal motility, obsessive-compulsive behaviors, and mood disorders.
- **simple carbohydrates (sugars):** monosaccharides and disaccharides.
- **soluble fibers:** indigestible food components that dissolve in water to form a gel. An example is pectin from fruit, which is used to thicken jellies.
- **starches:** plant polysaccharides composed of glucose.
- **sucrase:** an enzyme that hydrolyzes sucrose.
- **sucrose:** a disaccharide composed of glucose and fructose; commonly known as table sugar, beet sugar, or cane sugar. Sucrose also occurs in many fruits and some vegetables and grains.
- **type 1 diabetes:** the less common type of diabetes in which the pancreas fails to produce insulin.
- **type 2 diabetes:** the more common type of diabetes in which the cells fail to respond to insulin.
- **viscous:** a gel-like consistency.

Sample Test Questions

Select the best answer for each question.

1. Carbohydrates appear in virtually all _____ foods.
    a. plant
    b. animal
    c. health
    d. protein

2. Carbohydrates are made of the following 3 atoms:
    a. carbon, oxygen, and nitrogen.
    b. carbon, oxygen, and hydrogen.
    c. nitrogen, oxygen, and hydrogen.
    d. cadmium, oxygen, and hydrogen.

3. Which of the following compounds is a monosaccharide?
    a. sucrose
    b. fructose
    c. maltose
    d. lactose
    e. pectin

4. Which of the following are disaccharides?
    a. glucose, maltose
    b. fructose, sucrose
    c. galactose, lactose
    d. maltose, sucrose

5. Fructose is the sweetest of sugars because:
    a. its chemical structure delivers a sweet sensation.
    b. its chemical formula resembles glucose.
    c. it provides more kcalories than other sugars.
    d. it occurs naturally in honey.

6. A hydrolysis reaction can:
    a. bond two monosaccharides to form a disaccharide.
    b. split a disaccharide to form two monosaccharides.
    c. form a molecule of water.
    d. a and c
    e. b and c

7. Which reaction links two monosaccharides together?
    a. disaccharide   c. absorption
    b. hydrolysis   d. condensation

8. Disaccharides include:
    a. glucose.   d. sucrose.
    b. maltose.   e. b and d
    c. glycogen.

9. The principal carbohydrate of milk is:
    a. lactose.   c. maltose.
    b. sucrose.   d. glycogen.

10. Fruits are usually sweet because they contain:
    a. fiber.   c. simple sugars.
    b. complex carbohydrates.   d. fats.

11. An animal polysaccharide composed of glucose is called:
    a. fiber.   c. glycogen.
    b. dextrins.   d. an enzyme.

12. The difference between glycogen and starch is:
    a. the bonds in starch are fatty acids.
    b. the bonds in starch are single.
    c. the bonds in glycogen are not hydrolyzed by human enzymes.
    d. the glucose units are linked together differently.

13. Starch is made up of many glucose units bonded together.
    a. true   b. false

14. Fibers that do not dissolve in water are:
    a. soluble fibers.   c. fermentable fibers.
    b. viscous fibers.   d. insoluble fibers.

15. Which of these starches escape digestion and absorption in the small intestine?
    a. functional starches   c. resistant starches
    b. phytic starches   d. plant starches

16. Most carbohydrate absorption occurs in the:
    a. mouth.   d. large intestine.
    b. stomach.   e. b and c
    c. small intestine.

17. An enzyme that hydrolyzes sucrose is:
   a. glucose.
   b. lactase.
   c. pectins.
   d. sucrase.

18. Carbohydrate digestion occurs in:
   a. the stomach and small intestine.
   b. the mouth and small intestine.
   c. the stomach and colon.
   d. the mouth and pancreas.

19. A condition that results from an inability to digest lactose is called:
   a. lactose deficiency.
   b. lactose intolerance.
   c. hypoglycemia.
   d. lactase intolerance.

20. The main function of carbohydrate in the body is to:
   a. furnish the body with energy.
   b. provide materials for synthesizing cell walls.
   c. synthesize fat.
   d. insulate the body to prevent heat loss.

21. When blood glucose levels fall, the liver:
   a. combines excess glucose molecules.
   b. stores glucose as glycogen.
   c. dismantles stored glycogen.
   d. combines glucose to form molecules of fat.

22. The conversion of protein to glucose is called:
   a. ketosis
   b. protein-sparing action
   c. glycoproteins
   d. gluconeogenesis

23. Which of the following is true regarding ketosis?
   a. It is a desirable state for weight maintenance.
   b. It is a condition that disturbs the body's acid-base balance.
   c. It occurs when people eat too many kcalories.
   d. It results from the body's breakdown of protein.

24. What hormone signals the release of glucose out of storage?
   a. glucagon
   b. insulin
   c. testosterone
   d. serotonin

25. What condition results in the body's cells failing to respond to insulin?
   a. hyperlipidemia
   b. hypoglycemia
   c. type 1 diabetes
   d. type 2 diabetes

26. The _____ refers to how quickly glucose is absorbed after a person eats and how high blood sugar rises.
   a. hypoglycemia
   b. hyperglycemia
   c. glycemic response
   d. glycemic load

27. Sugar causes this (these) harmful condition(s):
   a. addictions.
   b. diabetes.
   c. ulcers.
   d. dental caries.
   e. all of the above

28. Taken with ample fluids, fibers can help prevent the following disorder(s):
   a. hemorrhoids.
   b. appendicitis.
   c. hyperactivity.
   d. a and b
   e. b and c

29. Dietary fiber:
   a. raises blood cholesterol levels.
   b. is found in high-fat foods.
   c. causes diverticulosis.
   d. provides fullness and delays hunger.

30. Excess fiber can result in:
   a. abdominal discomfort.
   b. gas and diarrhea.
   c. dental caries.
   d. a and b
   e. a, b, and c

Short Answer Questions

1. Hormones involved in maintaining blood glucose are:

   a.   c.

   b.

2. The monosaccharides important in nutrition are:

   a.   c.

   b.

3. The disaccharides important in nutrition are:

   a.   c.

   b.

4. The polysaccharides important in nutrition are:

   a.   c.

   b.

Problem Solving

1. How many grams of carbohydrate are in the following meal?
   2 slices whole-wheat bread with
   1 oz. cheddar cheese
   1 pat butter
   ½ cup orange juice

2. How many kcalories do carbohydrates contribute to the meal above?

3. The meal provides 365 kcalories. What percentage of the kcalories is from carbohydrates?

4. If a person's energy requirement is 1500 kcalories per day, calculate the number of grams of carbohydrate necessary to provide 55% of kcalories from carbohydrate.

5. If a person's energy requirement is 2500 kcalories per day, calculate the number of grams of carbohydrate necessary to provide 55% of kcalories from carbohydrate.

6. If a person consumed 31 teaspoons of white sugar, how many kcalories did they ingest?

7. A person who consumes 10 oz. of regular carbonated soda is taking in how many kcalories?

8. A person whose energy needs are 2000 kcalories per day consumes 10 tablespoons of catsup. How many kcalories were provided by the catsup, and what percent of kcalories were from the catsup alone?

9. If you consume 1.5 cups of dry oat bran, how many grams of fiber does this provide?

10. If a label indicates that a product provides 20 g of total carbohydrate, 2 grams of dietary fiber, and 5 grams of sugar, how many grams of starch does it provide?

Figure Identification

Identify these structures and reactions.

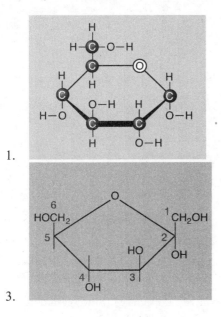

1.

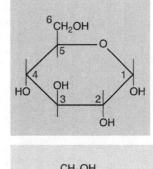

2.

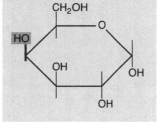

3.

4.

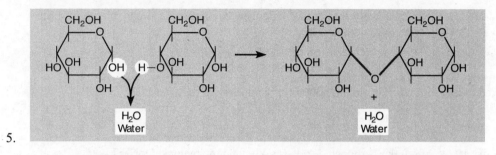

5.

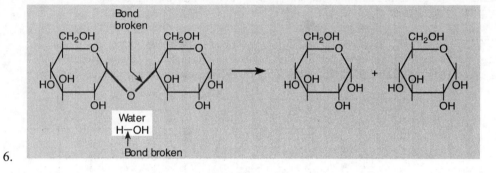

6.

Crossword Puzzle

Complete this crossword puzzle by Mary A. Wyandt, Ph.D., CHES.

Across	Down
3. an enzyme that hydrolyzes amylose	1. a non-nutrient component of plant seeds; also called phytate
5. to digest in the absence of oxygen	2. a sac or pouch that develops in the weakened areas of the intestinal wall
6. an abnormally low blood glucose concentration	
8. an enzyme that hydrolyzes maltose	4. an undesirably high concentration of ketone bodies in the blood and urine
9. a disorder of carbohydrate metabolism resulting from inadequate or ineffective insulin	7. a hormone secreted by special cells in the pancreas in response to (among other things) increased blood glucose concentration
10. a hormone secreted by special cells in the pancreas in response to low blood glucose concentration that elicits release of glucose from storage	

⑥ Chapter 4 Answer Key ⑧

Summing Up

1. carbohydrate	8. fructose	15. cells	22. pectin
2. starch	9. glucose	16. insulin	23. lignin
3. complex	10. starch	17. homeostasis	24. 14
4. polysaccharides	11. grains	18 45	25. 1000
5. monosaccharides	12. glycogen	19 65	
6. sucrose	13. liver	20. complex	
7. lactose	14. cellulose	21. exchange	

Chapter Study Questions

1 Simple—monosaccharides (glucose, fructose, galactose), disaccharides (sucrose, lactose, maltose). Complex—glycogen and starch.

2. A monosaccharide is a carbohydrate with a general structure of a ring composed of carbon, hydrogen and oxygen atoms. Monosaccharides important in nutrition are glucose, fructose, and galactose; disaccharides important in nutrition are sucrose (fructose + glucose), lactose (galactose + glucose), and maltose (2 glucoses). Nearly all plant foods contain glucose, and most plants (especially fruits and saps) contain fructose. Galactose is not found as such in foods. Sucrose occurs in many fruits and some vegetables and grains. Lactose is found in milk, and maltose is found in seeds.

3. Condensation combines two reactants to yield one product with the removal of water; hydrolysis splits one reactant into two products with the addition of water.

4. Polysaccharides are composed of many monosaccharides strung together. Important in nutrition are: glycogen, starch and the fibers. Starch and glycogen are similar in that they are both composed of glucose; they differ in the way their glucose units are linked together. Glycogen consists of many glucose molecules linked together in highly branched chains, while starch consists of many glucose molecules linked side by side. Fibers are different in that the bonds between their monosaccharides cannot be broken by human enzymes.

5. Salivary amylase enzymes in the mouth partially break down some of the starch before it reaches the intestine, pancreatic enzymes digest the starch to disaccharides in small intestine, disaccharidase enzymes on surface of intestinal wall cells split disaccharides to monosaccharides, monosaccharides enter capillary, capillary delivers monosaccharides to liver, liver converts galactose and fructose to glucose. In the mouth, fiber slows the process of eating and stimulates the flow of saliva; in the stomach, fiber delays gastric emptying; in the small intestine, it delays absorption of carbohydrates and fats, and can bind with minerals; in the large intestine, it attracts water that softens the stools.

6. It can be stored as glycogen; it can be used for energy; it can be converted to fat, when carbohydrate is available. Glucose can be used for energy, leaving protein available for its special functions.

7. Hormones are secreted in response to fluctuations in blood glucose. When blood glucose is too high, the pancreas releases insulin, resulting in the storage of glucose in the cells; when blood glucose is too low, the pancreas releases glucagon, resulting in the release of glucose into the blood.

8. Excess sugar can cause malnutrition if sugar displaces needed nutrients from the diet. It can contribute to obesity. Excess sugar can contribute to elevated blood lipids, and it can cause dental caries. The DRI recommend that no more than 25% of kcal intake should come from added sugars, whereas the World Health Organization suggests they should not exceed 10% of kcal intake.

9. They protect against heart disease, colon cancer, and diabetes, assist in weight control, improve large intestine function and health, lower blood cholesterol levels, and slow the rate of glucose absorption. 45 to 65% of total kcalories should come from carbohydrate; mostly from starch, some from fruits, vegetables, and milk. The DRI recommendation for fiber is 14 grams of fiber per 1000 kcalories consumed.

10. Fruits, vegetables, legumes, whole-grain breads and cereals.

Sample Test Questions

1. a (p. 101)	9. a (p. 105)	17. d (p. 108)	25. d (p. 115)
2. b (p. 101)	10. c (p. 103, 104)	18. b (p. 108)	26. c (p. 115)
3. b (p. 102)	11. c (p. 105)	19. b (p. 110)	27. d (p. 119-120)
4. d (p. 102)	12. d (p. 105-106)	20. a (p. 111)	28. d (p. 122)
5. a (p. 103)	13. a (p. 105)	21. c (p. 112)	29. d (p. 123)
6. b (p. 104)	14. d (p. 106)	22. d (p. 112)	30. d (p. 123)
7. d (p. 104)	15. c (p. 107)	23. b (p. 113)	
8. e (p. 102)	16. c (p. 108)	24. a (p. 113)	

Short Answer Questions

1. insulin; glucagon; epinephrine
2. glucose; fructose; galactose
3. maltose; sucrose; lactose
4. starch; glycogen; fiber

Problem Solving

1. 2 (15 g) + 0 + 0 + 15 g = 45 g
2. 4 kcal/g x 45 g = 180 kcal
3. 180 kcal divided by 365 kcal = 49%
4. 1500 kcal x .55 = 825 kcal divided by 4 = 206 grams
5. 2500 kcal x .55 = 1375 kcal divided by 4 = 344 grams
6. 31 teaspoons x 16 kcalories = 496
7. 1.5 oz = 20 kcalories
   20 divided by 1.5 = 13.3 kcalories per ounce
   10 oz. x 13.3 kcal per oz. = 133 kcalories
8. 10 tbsp x 20 kcalories = 200 kcalories
   200 divided by 2000 = 10% of kcalorie needs
9. .5 cups = 8 grams fiber
   1.5 divided by .5 = 3
   3 x 8 = 24 grams
10. 20 g total – 2 grams dietary fiber – 5 grams sugar = 13 g starch

Figure Identification

1. glucose
2. glucose
3. fructose
4. galactose
5. condensation reaction
6. hydrolysis reaction

Crossword Puzzle

1. phytic acid
2. diverticula
3. amylase
4. ketosis
5. ferment
6. hypoglycemia
7. insulin
8. maltase
9. diabetes
10. glucagon

☙ Chapter 5 – The Lipids: ❧
Triglycerides, Phospholipids, and Sterols

Chapter Outline

I. The Chemist's View of Fatty Acids and Triglycerides
   A. Fatty Acids
      1. The Length of the Carbon Chain
      2. The Degree of Unsaturation
      3. The Location of Double Bonds
   B. Triglycerides
   C. Degree of Unsaturation Revisited
      1. Firmness
      2. Stability
      3. Hydrogenation
      4. *Trans*-Fatty Acids
II. The Chemist's View of Phospholipids and Sterols
   A. Phospholipids
      1. Phospholipids in Foods
      2. Roles of Phospholipids
   B. Sterols
      1. Sterols in Foods
      2. Roles of Sterols
III. Digestion, Absorption, and Transport of Lipids
   A. Lipid Digestion
      1. In the Mouth
      2. In the Stomach
      3. In the Small Intestine
      4. Bile's Routes
   B. Lipid Absorption
   C. Lipid Transport
      1. Chylomicrons
      2. VLDL (Very-Low-Density Lipoproteins)
      3. LDL (Low-Density Lipoproteins)
      4. HDL (High-Density Lipoproteins)
      5. Health Implications
IV. Lipids in the Body
   A. Roles of Triglycerides
   B. Essential Fatty Acids
      1. Linoleic Acid and the Omega-6 Family
      2. Linolenic Acid and the Omega-3 Family

   3. Eicosanoids
   4. Fatty Acid Deficiencies
   C. A Preview of Lipid Metabolism
      1. Storing Fat as Fat
      2. Using Fat for Energy
V. Health Effects and Recommended Intakes of Lipids
   A. Health Effects of Lipids
      1. Heart Disease
      2. Risks from Saturated Fats
      3. Risks from *Trans* Fats
      4. Risks from Cholesterol
      5. Benefits from Monounsaturated Fats and Polyunsaturated Fats
      6. Benefits from Omega-3 Fats
      7. Balance Omega-6 and Omega-3 Intakes
      8. Cancer
      9. Obesity
   B. Recommended Intakes of Fat
   C. From Guidelines to Groceries
      1. Meats and Meat Alternates
      2. Milks and Milk Products
      3. Vegetables, Fruits, and Grains
      4. Invisible Fats
      5. Choose Wisely
      6. Fat Replacers
      7. Read Food Labels
VI. High-Fat Foods—Friend or Foe?
   A. Guidelines for Fat Intake
   B. High-Fat Foods and Heart Health
      1. Cook with Olive Oil
      2. Nibble on Nuts
      3. Feast on Fish
   C. High-Fat Foods and Heart Disease
      1. Limit Fatty Meats, Whole-Milk Products, and Tropical Oils
      2. Limit Hydrogenated Foods
   D. The Mediterranean Diet
   E. Conclusion

Summing Up

Lipids in the body, including (1)_____, oils, phospholipids, and sterols, function to maintain the health of the (2)_____ and hair; to protect body (3)_____ from heat, cold, and mechanical shock; and to provide a continuous (4)_____ supply. In foods, fats and oils act as a solvent for the fat-soluble (5)_____ and the compounds that give foods their

(6)_____ and aromas.

About 95 percent of the lipids in the diet are (7)_____; the phospholipids and sterols make up the remaining 5 percent. Triglycerides are composed of (8)_____ with (9)_____ fatty acids attached. The fatty acids may be classified as (10)_____, monounsaturated, or polyunsaturated.

During digestion, the triglycerides are emulsified by (11)_____ and then hydrolyzed by enzymes to (12)_____, glycerol, and fatty acids, which then pass into the intestinal cells. After absorption, all three classes of lipids are transported by (13)_____ in the body fluids. Cholesterol, a sterol, is synthesized in the body by the (14)_____. The body uses cholesterol in cell membranes and to make (15)_____ salts, hormones, and vitamin D.

Cholesterol from the liver may be transported to body tissues via the lipoproteins and may also be abnormally deposited in (16)_____ walls. A diet high in (17)_____ fat and cholesterol has been implicated as a causative factor in (18)_____. Most authorities recommend limiting excess fat, saturated fat, and cholesterol.

Most of the saturated fat found in the diet comes from (19)_____ and animal fats. Cholesterol is contributed by organ meats, shellfish, (20)_____, meats, and animal fats. (21)_____ products do not contain cholesterol. Vegetable and fish oils generally contain more (22)_____ fats than do animal fats. (23)_____ fats are fatty acids with hydrogens on opposite sides of the double bond. These fats (24)_____ LDL cholesterol and may increase inflammation and (25)_____ resistance.

Chapter Study Questions

1. Name the three classes of lipids found in the body and in foods. What are some of their functions in the body? What features do fats bring to foods?

2. What features distinguish fatty acids from each other?

3. What does the term "omega" mean with respect to fatty acids? Describe the roles of the omega fatty acids in disease prevention.

4. What are the differences between saturated, unsaturated, monounsaturated, and polyunsaturated fats? Describe the structure of a triglyceride.

5. What does hydrogenation do to fat? What are *trans*-fatty acids and how do they influence heart disease risk?

6. How do phospholipids differ from triglycerides in structure? How does cholesterol differ? How do these differences in structure affect function?

7. What roles do phospholipids play in the body? What roles does cholesterol play in the body?

8. Trace the steps in fat digestion, absorption, and transport. Describe the routes cholesterol takes in the body.

9. What do lipoproteins do? What are the differences among the chylomicrons, VLDL, LDL, and HDL?

10. Which of the fatty acids are essential? Name their chief dietary sources.

11. How does excessive fat intake influence health? What factors influence LDL, HDL, and total blood cholesterol?

12. What are the dietary recommendations regarding fat and cholesterol intake? List ways to reduce intake.

13. What is the Daily Value for fat? What does this number represent?

Chapter Glossary

- **adipose tissue:** the body's fat tissue; consists of masses of triglyceride-storing cells.
- **antioxidants:** in the body, compounds that protect others from oxidation by being oxidized themselves, thereby decreasing the adverse effects of free radicals on normal physiological functions.
- **arachidonic acid:** an omega-6 polyunsaturated fatty acid with 20 carbons and four double bonds; present in small amounts in meat and other animal products and synthesized in the body from linoleic acid.
- **artificial fats:** zero-energy fat replacers that are chemically synthesized to mimic the sensory and cooking qualities of naturally occurring fats, but are totally or partially resistant to digestion.
- **atherosclerosis:** a type of artery disease characterized by plaques (accumulations of lipid-containing material) on the inner walls of the arteries.
- **blood lipid profile:** results of blood tests that reveal a person's total cholesterol, triglycerides, and various lipoproteins.
- **cardiovascular disease (CVD):** a general term for all diseases of the heart and blood vessels. Atherosclerosis is the main cause of CVD. When the arteries that carry blood to the heart muscle become blocked, the heart suffers damage known as **coronary heart disease (CHD)**.
- **cholesterol:** one of the sterols containing four carbon ring structures with a carbon side chain.
- **choline:** a nitrogen-containing compound found in foods and made in the body from the amino acid methionine. Choline is part of the phospholipid lecithin and the neurotransmitter acetylcholine.
- **chylomicrons:** the class of lipoproteins that transport lipids from the intestinal cells to the rest of the body.
- **conjugated linoleic acid:** a collective term for several fatty acids that have the same chemical formula as linoleic acid (18 carbons, 20 double bonds) but with different configurations.
- **docosahexaenoic acid (DHA):** an omega-3 polyunsaturated fatty acid with 22 carbons and six double bonds; present in fish and synthesized in limited amounts in the body from linolenic acid.
- **eicosanoids:** derivatives of 20-carbon fatty acids; biologically active compounds that help to regulate blood pressure, blood clotting, and other body functions. They include *prostaglandins*, *thromboxanes*, and *leukotrienes*.
- **eicosapentaenoic acid (EPA):** an omega-3 polyunsaturated fatty acid with 20 carbons and five double bonds; present in fish and synthesized in limited amounts in the body from linolenic acid.
- **essential fatty acids:** fatty acids needed by the body, but not made by the body in amounts sufficient to meet physiological needs.
- **fat replacers:** ingredients that replace some or all of the functions of fat and may or may not provide energy.
- **fats:** lipids that are solid at room temperature (77° F or 25° C).
- **fatty acid:** an organic compound composed of a carbon chain with hydrogens attached and an acid group (COOH) at one end and a methyl group (CH$_3$) at the other end.
- **glycerol:** an alcohol composed of a three-carbon chain, which can serve as the backbone for a triglyceride.
- **HDL (high-density lipoprotein):** the type of lipoprotein that transports cholesterol back to the liver from the cells; composed primarily of protein.
- **hormone-sensitive lipase:** an enzyme inside adipose cells that responds to the body's need for fuel by hydrolyzing triglycerides so that their parts (glycerol and fatty acids) escape into the general circulation and thus become available to other cells as fuel. The signals to which this enzyme responds include epinephrine and glucagon, which oppose insulin.
- **hydrogenation:** a chemical process by which hydrogens are added to monounsaturated or polyunsaturated fatty acids to reduce the number of double bonds, making the fats more saturated (solid) and more resistant to oxidation (protecting against rancidity). Hydrogenation produces *trans*-fatty acids.
- **hydrophilic:** a term referring to water-loving, or water-soluble, substances.

- **hydrophobic:** a term referring to water-fearing, or non-water-soluble, substances; also known as **lipophilic** (fat loving).
- **LDL (low-density lipoprotein):** the type of lipoprotein derived from very-low-density lipoproteins (VLDL) as VLDL triglycerides are removed and broken down; composed primarily of cholesterol.
- **lecithin (LESS-uh-thin):** one of the phospholipids. Both nature and the food industry use lecithin as an emulsifier to combine water-soluble and fat-soluble ingredients that do not ordinarily mix, such as water and oil.
- **linoleic acid:** an essential fatty acid with 18 carbons and two double bonds.
- **linolenic acid:** an essential fatty acid with 18 carbons and three double bonds.
- **lipids:** a family of compounds that includes triglycerides, phopholipids, and sterols. Lipids are characterized by their insolubility in water.
- **lipoprotein lipase (LPL):** an enzyme that hydrolyzes triglycerides passing by in the bloodstream and directs their parts into the cells, where they can be metabolized for energy or reassembled for storage.
- **lipoproteins:** clusters of lipids associated with proteins that serve as transport vehicles for lipids in the lymph and blood.
- **micelles:** tiny spherical complexes of emulsified fat that arise during digestion; most contain bile salts and the products of lipid digestion, including fatty acids, monoglycerides, and cholesterol.
- **monoglycerides:** molecules of glycerol with one fatty acid attached. A molecule of glycerol with two fatty acids attached is a **diglyceride**.
- **monounsaturated fatty acid:** a fatty acid that lacks two hydrogen atoms and has one double bond between carbons—for example, oleic acid.
- **oils:** lipids that are liquid at room temperature (77° F or 25° C).
- **olestra:** a synthetic fat made from sucrose and fatty acids that provides 0 kcalories per gram; also known as **sucrose polyester**.
- **omega:** the last letter of the Greek alphabet (ω), used by chemists to refer to the position of the first double bond from the methyl (CH3) end of a fatty acid.
- **omega-3 fatty acid:** a polyunsaturated fatty acid in which the first double bond is three carbons away from the methyl (CH_3) end of the carbon chain.
- **omega-6 fatty acid:** a polyunsaturated fatty acid in which the first double bond is six carbons from the methyl (CH_3) end of the carbon chain.
- **oxidation:** the process of a substance combining with oxygen; oxidation reactions involve the loss of electrons.
- **phospholipid:** a compound similar to a triglyceride but having a phosphate group (a phosphorus-containing salt) and choline (or another nitrogen-containing compound) in place of one of the fatty acids.
- **point of unsaturation:** the double bond of a fatty acid, where hydrogen atoms can easily be added to the structure.
- **polyunsaturated fatty acid (PUFA):** a fatty acid that lacks four or more hydrogen atoms and has two or more double bonds between carbons—for example, linoleic acid (two double bonds) and linolenic acid (three double bonds).
- **saturated fatty acid:** a fatty acid carrying the maximum possible number of hydrogen atoms—for example, stearic acid.
- **sterols:** compounds containing four carbon ring structures with any of a variety of side chains attached.
- *trans*-**fatty acids:** fatty acids with hydrogens on opposite sides of the double bond.
- **triglycerides:** the chief form of fat in the diet and the major storage form of fat in the body; composed of a molecule of glycerol with three fatty acids attached; also called **triacylglycerols**.
- **unsaturated fatty acid:** a fatty acid that lacks hydrogen atoms and has at least one double bond between carbons (includes monounsaturated and polyunsaturated fatty acids).
- **VLDL (very-low-density lipoprotein):** the type of lipoprotein made primarily by liver cells to transport lipids to various tissues in the body; composed primarily of triglycerides.

Sample Test Questions

Select the best answer for each question.

1. Fats belong to a larger chemical classification known as:
   - a. lipases.
   - b. lecithins.
   - c. labiles.
   - d. leucines.
   - e. lipids.

2. Of the lipids in foods, what percent are triglycerides?
   - a. 95%
   - b. 90%
   - c. 88%
   - b. 85%

3. Fatty acids and triglycerides are composed of which atoms?
   - a. carbon, hydrogen, nitrogen
   - b. carbon, nitrogen, helium
   - c. hydrogen, mercury, oxygen
   - d. carbon, hydrogen, oxygen

4. Triglycerides are composed of:
   - a. glycerol.
   - b. pyruvate.
   - c. fatty acids.
   - d. a and c
   - e. b and c

5. The position of the double bond nearest the methyl group is described by a(n) _____ number.
   - a. omega
   - b. alpha
   - c. delta
   - d. polyunsaturated

6. Oleic acid has one double bond in the carbon chain. This means it is classified as:
   - a. saturated.
   - b. monounsaturated.
   - c. diunsaturated.
   - d. polyunsaturated.

7. The degree of unsaturation influences fats in all of the following ways EXCEPT:
   - a. kcalorie content.
   - b. firmness.
   - c. stability.
   - d. oxidation.

8. At room temperature, which type of fat is usually solid?
   - a. unsaturated fat
   - b. saturated fat
   - c. polyunsaturated fatty acid
   - d. oils

9. A disadvantage of hydrogenation is:
   - a. the fats become resistant to oxidation.
   - b. *cis*-fatty acids result.
   - c. *trans*-fatty acids result.
   - d. a and b
   - e. b and c

10. Hydrogenation of fat renders it:
    - a. more nutritious.
    - b. more susceptible to oxidation and rancidity.
    - c. less susceptible to oxidation and rancidity.
    - d. less saturated, thus less stable.

11. The manufacturing process that makes polyunsaturated fats more solid is called:
    - a. centrifugation.
    - b. hydrogenation.
    - c. emulsification.
    - d. modification.
    - e. solidification.

12. The dispersion and stabilization of fat droplets in a watery solution is:
    a. hydrogenation.
    b. emulsification.
    c. saturation.
    d. precipitation.

13. Lecithins are:
    a. carbohydrates.
    b. triglycerides.
    c. proteins.
    d. phospholipids.

14. This fat is a compound with a multiple-ring structure:
    a. phospholipid.
    b. sterol.
    c. glycerol.
    d. fatty acid.

15. Cholesterol that is made in the body is:
    a. undesirable.
    b. bile acids.
    c. exogenous.
    d. endogenous.

16.   Bile is made from:
    a. glucose.
    b. cholesterol.
    c. vitamin A.
    d. prostaglandins.

17. A lipoprotein may be described as:
    a. a molecule made up of various amino acids.
    b. a triglyceride made of glycerol and three fatty acids.
    c. a cluster of lipids associated with proteins.
    d. undigested fat circulating in the bloodstream.

18. The process of lipid absorption involves some molecules merging into spherical complexes called:
    a. micelles
    b. bile
    c. intestinal cells
    d. fat

19. The largest and least dense of the lipoproteins are the:
    a. chylomicrons.
    b. LDL.
    c. VLDL.
    d. HDL.

20. The lipoprotein thought to protect against coronary heart disease is:
    a. LDL.
    b. VLD.
    c. IDL.
    d. HDL.
    e. the chylomicron.

21. The primary member of the omega-3 family is:
    a. lecithin.
    b. eicosanoids.
    c. linolenic acid.
    d. linoleic acid.

22. Which of the following is not lost when fat or oil is removed from a food?
    a. flavor
    b. aroma
    c. kcalories
    d. vitamin E
    e. vitamin C

23. Which enzyme hydrolyzes triglycerides from lipoproteins?
    a. lipoprotein lipase (LPL)
    b. hormone-sensitive lipase (HSL)
    c. insulin lipase (IL)
    d. lipoprotein protease (LPP)

24. Fat supplies what percent of the body's ongoing energy needs during rest?
    a. 30%
    b. 40%
    c. 50%
    d. 60%

25. The type of fat in food that influences blood cholesterol the most is:
    a. cholesterol.
    b. saturated.
    c. unsaturated.
    d. polyunsaturated.

26. Which of the following foods contains an appreciable amount of *trans* fat?
    a. skim milk
    b. bread
    c. cakes and cookies
    d. vegetables
    e. fruits

27. Which contains the most saturated fat?
    a. sunflower seed oil
    b. corn oil
    c. lard
    d. olive oil

28. Which of the following foods contains no cholesterol?
    a. beef steak
    b. bacon
    c. ice cream
    d. baked potato

29. A change from whole milk to nonfat milk would:
    a. decrease the amount of fat and kcalories.
    b. increase the number of kcalories.
    c. increase satiety value of the meal.
    d. decrease the amount of essential amino acids.

30. Fat replacers that offer the desirable qualities of fat without the kcalories are:
    a. triglycerides.
    b. chemical fats.
    c. artificial fats.
    d. hydrogenated fats.

Short Answer Questions

1. The lipids include:

   a.   c.

   b.

2. The roles of body fat are:

   a.   c.

   b.   d.

3. The roles of food fat are:

   a.   c.

   b.   d.

4. Triglycerides are made of:

   a.   b.

5. The lipoproteins are:

   a.   c.

   b.   d.

Problem Solving

1. Calculate the number of kcalories from fat for a person who requires 2,000 kcalories per day and who consumes 35% of kcalories from fat.

2. In the above problem, how many grams of fat would be consumed?

3. Your kcalorie needs are 2200 kcalories per day. How many kcalories should come from saturated fat?

4. In the above problem, how many grams of saturated fat would be consumed if you ate the maximum recommended?

5. Your kcalorie needs are 1700 kcalories per day and you are striving to consume a diet that provides 30% of your kcalories from fat. How many kcalories from fat must you consume?

6. In the above problem, how many grams of fat would you ingest?

7. You have prepared a milkshake that includes 2 cups of whole milk with ½ cup strawberries. How many grams of fat are in this product? How many kcalories from fat are in the milkshake?

8. Regarding the above question, what percentage of the Daily Value will this milkshake provide?

9. You are considering eating a piece of pecan pie that provides 400 kcalories from fat. How many grams of fat does this dessert provide?

10. If your energy requirement is 1600 kcalories per day, what percent of your fat allotment will the pie provide, assuming you are trying to eat 35% of total kcalories from fat each day?

Figure Identification

Identify these chemical structures.

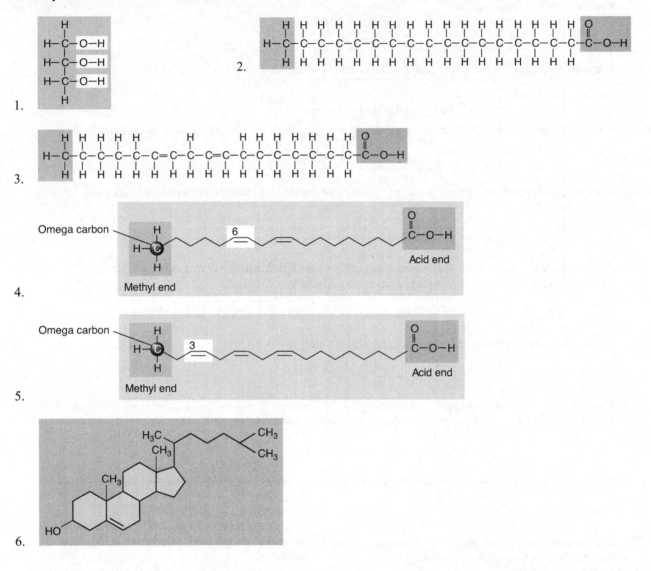

1.

2.

3.

4.

5.

6.

Crossword Puzzle

Complete this crossword puzzle by Mary A. Wyandt, Ph.D., CHES.

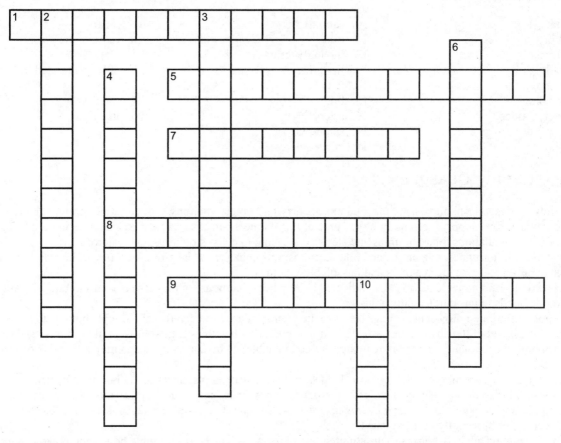

Across:	Down:
1. One of the sterols containing a four-carbon ring structure with a carbon side chain.	2. A term referring to water-loving, or water-soluble, substances.
5. Results of blood tests that reveal a person's total cholesterol, triglycerides, and various lipoproteins.	3. The chief form of fat in the diet and the major storage form of fat in the body; composed of a molecule of glycerol with three fatty acids attached.
7. An alcohol composed of a three-carbon chain, which can serve as the backbone for a triglyceride.	4. A compound similar to a triglyceride but having a phosphate group and choline in place of one of the fatty acids.
8. High blood pressure.	6. Derivatives of 20-carbon fatty acids; biologically active compounds that help to regulate blood pressure, blood clotting, and other body functions.
9. Clusters of lipids associated with proteins that serve as transport vehicles for lipids in the lymph and blood.	10. The last letter of the Greek alphabet, used by chemists to refer to the position of the first double bond from the methyl end of a fatty acid.

⑤ Chapter 5 Answer Key ⑨

Summing Up

1. fats
2. skin
3. organs
4. fuel
5. vitamins
6. flavors
7. triglycerides
8. glycerol
9. 3

10. saturated
11. bile
12. monoglycerides
13. lipoproteins
14. liver
15. bile
16. artery
17. saturated
18. atherosclerosis

19. meat
20. eggs
21. Vegetable
22. polyunsaturated
23. *Trans*
24. raise
25. insulin

Chapter Study Questions

1. Triglycerides, phospholipids, sterols. Functions: carry fat-soluble vitamins, induce satiety, provide body with a continuous food supply, keep body warm, protect it from mechanical shock, serve as starting materials for hormonal regulations; phospholipids and sterols contribute to cells' structures; cholesterol serves as raw material for hormones, vitamin D, and bile. Features: fats enhance foods' aroma and flavor, increase palatability, and provide kcalories and fat-soluble vitamins.

2. Essentiality and nonessentiality; size indicated by number of carbons; saturated versus unsaturated, location of double bonds in unsaturated fatty acids.

3. In polyunsaturated fatty acids, omega refers to the relative place in which the first double bond is located from the methyl end of the chain. These structures are important to health as they are essential nutrients used to make hormone-like substances that play regulatory roles in the body. Omega-3 fatty acids may lower risk of some cancers and heart disease.

4. Saturated fats have all their carbon atoms loaded with hydrogen atoms; unsaturated fats have hydrogens removed; monounsaturated fats have two hydrogens removed and one double bond; polyunsaturated fats have multiple double bonds and several hydrogens missing. Structurally, a triglyceride consists of a "backbone" of glycerol with three fatty acids attached.

5. Hydrogenation adds hydrogens to unsaturated fats to reduce the number of double bonds and make fats more saturated and resistant to oxidation. Hydrogenation can increase a product's shelf life and add a desirable texture. A *trans*-fatty acid is one in which the hydrogens next to the double bonds are on opposite sides of the carbon chain; *trans*-fatty acids are thought to increase heart disease risk. *Trans*-fatty acids are associated with heart disease risk and experts advise that intakes of these fatty acids be limited.

6. Phospholipids have a choline or other phosphorus-containing acid in place of one of the fatty acids, enabling them to function as emulsifiers in the body. The phospholipid allows the fatty acids to dissolve in water. Cholesterol is a sterol, and its C, H, and O atoms are arranged in rings. Cholesterol in the body can serve as starting material for many important body compounds.

7. Phospholipids are important parts of cell membranes, they help lipids move back and forth across the cell membranes into the watery fluids on both sides, and they enable fat-soluble vitamins and hormones to pass easily in and out of cells. Cholesterol's vital roles are many, including providing energy and serving as part of cell membranes, bile acids, sex hormones, adrenal hormones, and vitamin D.

8. Bile emulsifies fats, allowing the enzymes to gain access to the fat for digestion; products of lipid digestion are packaged with protein for transport. Cholesterol shuttles back and forth between the liver and the body cells in lipoproteins, and it also visits the intestinal tract in the form of bile.

9. Composed of triglycerides, cholesterol, phospholipids, and proteins, lipoproteins transport lipids in the body. Chylomicrons are the largest of the lipoproteins, formed in the intestinal wall following fat absorption, and they contain mostly triglycerides. VLDL are made in the liver and contain mostly triglycerides; LDL contain few triglycerides but are about half cholesterol. HDL are about half protein and transport cholesterol back to the liver.

10. Linolenic acid and linoleic acid; vegetable oils and meats, grains, seeds, nuts, leafy vegetables, fish.

11. Excessive fat intake can contribute to elevated blood cholesterol and other blood lipids (therefore heart disease), obesity, and cancer. Some saturated fats raise total cholesterol and LDL; *trans*-fatty acids raise LDL and lower HDL. Monounsaturated fats lower LDL without lowering HDL.

12. To consume a diet that is low in saturated fat, *trans* fat, and cholesterol (300 mg), limit total fat intake to 20 to 35% of daily energy from fat; consume 5 to 10 percent of daily energy from linoleic acid and 0.6 to 1.2 percent from linolenic acid. To reduce fat intake select lean meats and nonfat milk, eat plenty of vegetables, fruits, and grains, use fats and oils sparingly, look for invisible fat, and read food labels.

13. 65 grams; Daily Value for a person consuming 2000 kcalories per day. For a person who consumes 2000 kcalories a day, this number represents the recommended number of grams of fat for a day.

Sample Test Questions

1. e (p. 139)	9. c (p. 143-144)	17. c (p. 150)	25. b (p. 157)
2. a (p. 139)	10. c (p. 143-144)	18. a (p. 149)	26. c (p. 157)
3. d (p. 139)	11. b (p. 143)	19. a (p. 151)	27. c (p. 144, 157)
4. d (p. 139)	12. b (p. 145, 149)	20. d (p. 152)	28. d (p. 146, 157)
5. a (p. 141)	13. d (p. 145)	21. c (p. 154)	29. a (p. 162)
6. b (p. 141)	14. b (p. 146)	22. e (p. 161)	30. c (p. 164)
7. a (p. 142-143)	15. d (p. 146)	23. a (p. 155)	
8. b (p. 142)	16. b (p. 147, 149)	24. d (p. 156)	

Short Answer Questions

1. triglycerides; phospholipids; sterols
2. maintain cell structure; protect organs; provide energy; protect lean tissue from depletion
3. provide palatability; provide satiety; deliver fat-soluble vitamins; provide kcalories
4. glycerol; 3 fatty acids
5. chylomicron; VLDL; LDL; HDL

Problem Solving

1. 2,000 kcal/day x .35 = 700 kcalories
2. 700 divided by 9 = 78
3. 2200 divided by 10 = 220 kcalories or less
4. 220 divided by 9 = 24 grams
5. 1700 x .30 = 510 kcalories from fat
6. 510 divided by 9 = 57 grams of fat
7. 2 cups x 8 grams fat = 16 grams of fat
   16 x 9 = 144 kcalories from fat
8. 16 grams divided by 65 grams fat = 25% of Daily Value for fat
9. 400 divided by 9 = 44 grams of fat
10. 1600 x .35 = 560 kcalories per day from fat
    560 divided by 9 = 62 total grams of fat
    44 divided by 62 = 71 percent from the pie

Figure Identification

1. glycerol
2. stearic acid
3. a polyunsaturated fatty acid (linoleic acid)
4. linoleic acid (an omega-6 fatty acid)
5. linolenic acid (an omega-3 fatty acid)
6. cholesterol

Crossword Puzzle

1. cholesterol
2. hydrophilic
3. triglycerides
4. phospholipids
5. lipid profile
6. eicosanoids
7. glycerol
8. hypertension
9. lipoprotein
10. omega

☉ Chapter 6 – Protein: Amino Acids ☉

Chapter Outline

I.   The Chemist's View of Proteins
   A. Amino Acids
      1. Unique Side Groups
      2. Nonessential Amino Acids
      3. Essential Amino Acids
      4. Conditionally Essential Amino Acids
   B. Proteins
      1. Amino Acid Chains
      2. Amino Acid Sequences
      3. Protein Shapes
      4. Protein Functions
      5. Protein Denaturation
II. Digestion and Absorption of Protein
   A. Protein Digestion
      1. In the Stomach
      2. In the Small Intestine
   B. Protein Absorption
III. Proteins in the Body
   A. Protein Synthesis
      1. Delivering the Instructions
      2. Lining Up the Amino Acids
      3. Sequencing Errors
      4. Nutrients and Gene Expression
   B. Roles of Proteins
      1. As Building Materials for Growth and Maintenance
      2. As Enzymes
      3. As Hormones
      4. As Regulators of Fluid Balance
      5. As Acid-Base Regulators
      6. As Transporters
      7. As Antibodies
      8. As a Source of Energy and Glucose
      9. Other Roles
   C. A Preview of Protein Metabolism
      1. Protein Turnover and the Amino Acid Pool
      2. Nitrogen Balance
      3. Using Amino Acids to Make Proteins or Nonessential Amino Acids
      4. Using Amino Acids to Make Other Compounds

      5. Using Amino Acids for Energy and Glucose
      6. Deaminating Amino Acids
      7. Using Amino Acids to Make Fat
IV. Protein in Foods
   A. Protein Quality
      1. Digestibility
      2. Amino Acid Composition
      3. Reference Protein
      4. High-Quality Proteins
      5. Complementary Proteins
   B. Protein Regulations for Food Labels
V.   Health Effects and Recommended Intakes of Protein
   A. Protein-Energy Malnutrition
      1. Classifying PEM
      2. Marasmus
      3. Kwashiorkor
      4. Marasmus-Kwashiorkor Mix
      5. Infections
      6. Rehabilitation
   B. Health Effects of Protein
      1. Heart Disease
      2. Cancer
      3. Adult Bone Loss (Osteoporosis)
      4. Weight Control
      5. Kidney Disease
   C. Recommended Intakes of Protein
      1. Protein RDA
      2. Adequate Energy
      3. Protein in Abundance
   D. Protein and Amino Acid Supplements
      1. Protein Powders
      2. Amino Acid Supplements
VI. Nutritional Genomics
   A. A Genomics Primer
   B. Genetic Variation and Disease
      1. Single-Gene Disorders
      2. Multigene Disorders
   C. Clinical Concerns

Summing Up

Proteins are composed of (1)_____ acids. Nonessential amino acids can be

(2)_____ in the body from carbohydrate derivatives and an amino nitrogen source.

(3)_____ amino acids cannot be synthesized in the body, or cannot be made in amounts sufficient to meet physiological need. The major role of dietary protein is to supply amino acids for the synthesis of (4)_____ needed in the body, although dietary protein can also serve as an (5)_____ source. The body's oxygen carrier, (6)_____, and the muscle's oxygen reservoir, myoglobin, are also proteins.

The study of the body's proteins is called (7)_____. The human genome is the full set of chromosomes, including all of the (8)_____ and associated DNA. The instructions for making every protein in a person's body are transmitted by way of the (9)_____ information received at conception. When messenger RNA is made from a template of DNA, (10)_____ occurs. The process of messenger RNA directing the sequence of amino acids and synthesis of proteins is (11)_____. Cells can regulate gene (12)_____ to make the type of proteins needed.

Proteins act as (13)_____ and they help regulate water balance and (14)_____ balance. (15)_____ and some hormones are made of proteins. Other body proteins include blood (16)_____ factors, collagen, and the light-sensitive (17)_____ of the retina. Maintenance and repair of body tissue and growth of new tissue require a continual supply of (18)_____ acids to synthesize proteins.

A (19)_____ protein supplies all the essential amino acids; a high-quality protein not only supplies them, but also provides them in the appropriate (20)_____. (21)_____ protein sources are generally of higher quality than vegetable protein sources, but diets composed of plant foods provide plenty of protein, as long as a variety of nutrient-dense foods supply a sufficient energy intake.

Kwashiorkor is a disease in which (22)_____ is lacking. Protein deficiency combined with inadequate food (23)_____ is marasmus. These deficiencies are called protein-energy (24)_____ (PEM) and are a worldwide malnutrition problem. Excessive intakes of protein offer no benefit; in fact, overconsumption of protein-rich foods may incur health (25)_____. Recommended protein intake is (26)_____ grams of protein per kilogram of body weight each day. Normal healthy people never need protein or (27)_____ supplements.

Chapter Study Questions

1. How does the chemical structure of proteins differ from the structures of carbohydrates and fats?

2. Describe the structure of amino acids, and explain how their sequence in proteins affects the proteins' shapes. What are the essential amino acids?

70

3. Describe protein digestion and absorption.

4. Describe protein synthesis.

5. Describe some of the roles proteins play in the human body.

6. What are enzymes? What roles do they play in chemical reactions? Describe the differences between enzymes and hormones.

7. How does the body use amino acids? What is deamination? Define nitrogen balance. What conditions are associated with zero, positive, and negative balance?

8. What factors affect the quality of dietary protein? What is a complete protein?

9. How can vegetarians meet their protein needs without eating meat?

10. What are the health consequences of ingesting inadequate protein and energy? Describe marasmus and kwashiorkor. How can the two conditions be distinguished, and in what ways do they overlap?

11. How might protein excess, or the type of protein eaten, influence health?

12. What factors are considered in establishing recommended protein intakes?

13. What are the benefits and risks of taking protein and amino acid supplements?

Chapter Glossary

- **acidosis:** above-normal acidity in the blood and body fluids.
- **acids:** compounds that release hydrogen ions in a solution.
- **acute PEM:** protein-energy malnutrition caused by recent severe food restriction; characterized in children by thinness for height (wasting).
- **alkalosis:** above-normal alkalinity (base) in the blood and body fluids.
- **amino acid pool:** the supply of amino acids derived from either food proteins or body proteins that collect in the cells and circulating blood and stand ready to be incorporated in proteins and other compounds or used for energy.
- **amino acids:** building blocks of proteins. Each contains an amino group, an acid group, a hydrogen atom, and a distinctive side group, all attached to a central carbon atom.
- **antibodies:** large proteins of the blood and body fluids, produced by the immune system in response to the invasion of the body by foreign molecules (usually proteins called *antigens*). Antibodies combine with and inactivate the foreign invaders, thus protecting the body.
- **antigens:** substances that elicit the formation of antibodies or an inflammation reaction from the immune system. A bacterium, a virus, a toxin, and a protein in food that causes allergy are all examples of antigens.
- **bases:** compounds that accept hydrogen ions in a solution.
- **branched-chain amino acids:** the essential amino acids leucine, isoleucine, and valine, which are present in large amounts in skeletal muscle tissue; falsely promoted as fuel for exercising muscles.
- **buffers:** compounds that help keep a solution's acidity or alkalinity constant.
- **chronic PEM:** protein-energy malnutrition caused by long-term food deprivation; characterized in children by short height for age (stunting).
- **collagen:** the protein from which connective tissues such as scars, tendons, ligaments, and the foundations of bones and teeth are made.
- **complementary proteins:** two or more dietary proteins whose amino acid assortments complement each other in such a way that the essential amino acids missing from one are supplied by the other.
- **conditionally essential amino acid:** an amino acid that is normally nonessential, but must be supplied by the diet in special circumstances when the need for it exceeds the body's ability to produce it.
- **deamination:** removal of the amino (NH_2) group from a compound such as an amino acid.
- **denaturation:** the change in a protein's shape and consequent loss of its function brought about by heat, agitation, acid, base, alcohol, heavy metals, or other agents.
- **dipeptide:** two amino acids bonded together.
- **dysentery:** an infection of the digestive tract that causes diarrhea.
- **edema:** the swelling of body tissue caused by excessive amounts of fluid in the interstitial spaces; seen in protein deficiency (among other conditions).
- **endogenous:** describes an amino acid (or protein) that derives from within the body.
- **enzymes:** proteins that facilitate chemical reactions without being changed in the process; protein catalysts.
- **essential amino acids:** amino acids that the body cannot synthesize in amounts sufficient to meet physiological needs.
- **exogenous:** describes an amino acid (or protein) that derives from foods

- **fluid balance:** maintenance of the proper types and amounts of fluid in each compartment of the body fluids.
- **gene expression:** the process by which a cell converts the genetic code into RNA and protein.
- **hemoglobin:** the globular protein of the red blood cells that carries oxygen from the lungs to the cells throughout the body.
- **high-quality proteins:** dietary proteins containing all the essential amino acids in relatively the same amounts that human beings require. They may also contain nonessential amino acids.
- **immunity:** the body's ability to defend itself against diseases.
- **kwashiorkor:** a form of PEM that results either from inadequate protein intake or, more commonly, from infections.
- **limiting amino acid:** the essential amino acid found in the shortest supply relative to the amounts needed for protein synthesis in the body.
- **marasmus:** a form of PEM that results from a severe deprivation, or impaired absorption, of energy, protein, vitamins, and minerals.
- **matrix:** the basic substance that gives form to a developing structure; in the body, the formative cells from which teeth and bones grow.
- **neurotransmitters:** chemicals that are released at the end of a nerve cell when a nerve impulse arrives there. They diffuse across the gap to the next cell and alter the membrane of that second cell to either inhibit or excite it.
- **nitrogen balance:** the amount of nitrogen consumed (N in) as compared with the amount of nitrogen excreted (N out) in a given period of time.
- **nonessential amino acids:** amino acids that the body can synthesize.
- **oligopeptide:** a string of four to nine amino acids bonded together.
- **pepsin:** a gastric enzyme that hydrolyzes protein.
- **pepsinogen:** the inactive form of pepsin which is activated by hydrochloric acid in the stomach.
- **peptidase:** a digestive enzyme that hydrolyzes peptide bonds.
- **peptide bond:** a bond that connects the acid end of one amino acid with the amino end of another, forming a link in a protein chain.
- **polypeptide:** many (ten or more) amino acids bonded together.
- **protease:** an enzyme that hydrolyze protein.
- **protein digestibility:** a measure of the amount of amino acids absorbed from a given protein intake.
- **protein-digestibility-corrected amino acid score (PDCAAS):** a measure of protein quality assessed by comparing the amino acid balance of a food protein with the amino acid requirements of preschool-aged children and then correcting for the true digestibility of protein.
- **protein-energy malnutrition (PEM)**, also called **protein-kcalorie malnutrition (PCM):** a deficiency of protein, energy, or both, including kwashiorkor, marasmus, and instances in which they overlap.
- **proteins:** compounds composed of carbon, hydrogen, oxygen, and nitrogen atoms, arranged into amino acids linked in a chain. Some amino acids also contain sulfur atoms.
- **protein turnover:** the degradation and synthesis of protein.
- **reference protein:** a standard against which to measure the quality of other proteins.
- **sickle-cell anemia:** a hereditary form of anemia characterized by abnormal sickle- or crescent-shaped red blood cells.
- **tripeptide:** three amino acids bonded together.
- **whey protein:** a by-product of cheese production; falsely promoted as increasing muscle mass. Whey is the watery part of milk that separates from the curds.

Sample Test Questions

Select the best answer for each question.

1. Proteins differ from the other energy nutrients in that they contain:
   a. glycerol.
   b. carbon.
   c. fatty acids.
   d. nitrogen.

2. A polypeptide is:
    a. many amino acids linked together by peptide bonds.
    b. formation of a helix by adenine, thymine, guanine and cystosine.
    c. the primary structure of an amino acid.
    d. composed of glucose and amino acids.

3. The structures of amino acids differ in that:
    a. some do not contain nitrogen.
    b. some do not contain carbon.
    c. they each have a different side chain.
    d. they each have different acid groups.

4. When an enzyme is used in a chemical reaction in the body, it:
    a. forms ATP.
    b. remains unchanged.
    c. is degraded into a substance containing less chemical energy.
    d. only breaks down substances.

5. Proteins maintain acid-base balance in the body by:
    a. acting as buffers.
    b. secreting chloride ions.
    c. secreting acids.
    d. tying up excess sodium ions.

6.-11. Match the following:

6. _____ enzyme   a. elicits the formation of antibodies
7. _____ antibody   b. protein catalyst
8. _____ hormone   c. inactivates foreign agents
9. _____ opsin   d. a chemical messenger
10. _____ antigen   e. helps form a blood clot
11. _____ fibrin   f. visual pigment protein

12. Methionine, threonine, and tryptophan are names of:
    a. proteins.
    b. fatty acids.
    c. essential amino acids.
    d. lipids.
    e. nonessential amino acids.

13. If an essential amino acid required for formation of a certain enzyme is missing in the diet:
    a. another amino acid will be substituted in its place so the enzyme can be made.
    b. synthesis of the enzyme will stop.
    c. the partially synthesized enzyme will be stored in the adipose tissue until the missing amino acid is supplied in the diet.
    d. the amino acid will be made from glucose.

14. The study of the body's proteins is called:
    a. digestible proteins.
    b. amino acid scoring.
    c. meat protein.
    d. proteomics.

15. The process of messenger RNA being made from a template of DNA is:
    a. transcription.
    b. translation.
    c. *trans* fatty acid.
    d. messenger RNA replication.

16. A body of knowledge is filed in this part of the cell:
    a. protein strand.
    b. ribosomes.
    c. RNA.
    d. DNA.

17. The making of specific types of proteins in certain amounts and at certain rates is called:
    a. epigenetics.
    b. replication.
    c. gene expression.
    d. translation.

18. An excess of interstitial fluid can cause:
    a. intravascular storage.
    b. edema.
    c. extracellular imbalance.
    d. protein reabsorption.

19. The body can normally detect invading disease-causing substances that are called:
    a. antibodies.
    b. immunity.
    c. a matrix.
    d. antigens.

20. If the body degrades more protein than it synthesizes and loses protein, nitrogen balance becomes:
    a. nitrogen output.
    b. negative.
    c. protein efficiency ratio.
    d. positive.

21. When amino acids are used for energy, they are broken down. This is called:
    a. deamination.
    b. amination.
    c. neurotransmitter formation.
    d. positive nitrogen balance.

22. The quality of a food protein:
    a. is judged by comparing its composition to amino acid requirements of preschool-aged children.
    b. is partially determined by its amino acid composition.
    c. is based on comparison with a reference protein.
    d. a and b
    e. b and c
    f. a, b and c

23. These foods tend to contain high-quality proteins:
    a. egg, meat, fish, corn, and seeds.
    b. egg, meat, poultry, soy, and milk.
    c. egg, meat, fish, gelatin, and grains.
    d. meat, milk, cheese, corn, and spinach.

24. The most widespread form of malnutrition in developing countries is:
    a. anorexia.
    b. bulimia.
    c. protein-energy malnutrition.
    d. iron-deficiency anemia.
    e. vitamin C deficiency.

25. Marasmus may be described as a nutritional deficiency disease with:
    a. severe deficiencies of protein and vitamins.
    b. severe deficiency of protein.
    c. severe deficiencies of protein, energy, vitamins, and minerals.
    d. excessive amounts of edema.

26. Excessive protein intake is a factor in the following conditions:
    a. heart disease, cancer, and osteoporosis.
    b. heart disease, anorexia, and cancer.
    c. heart disease, kidney disease, and bulimia.
    d. cancer, osteoporosis, and dental caries.

27. The protein RDA for adults is:
    a. 0.8 g per pound of body weight.
    b. 0.8 g per kg of body weight.
    c. 1.2 g per pound body weight.
    d. 1.5 g per kg body weight.

28. When establishing the protein RDA, the committee assumed:
    a. that people have unusual metabolic needs for protein.
    b. that the protein consumed will be of high quality.
    c. that protein will be consumed along with sufficient carbohydrate and fat.
    d. all of the above

29. Protein and amino acid supplements:
    a. can cure herpes.
    b. can facilitate weight loss.
    c. can strengthen fingernails.
    d. can be harmful.

30. Large doses of branched-chain amino acids may:
    a. lower plasma concentrations of ammonia.
    b. help athletes with performance.
    c. raise ammonia levels, which may be toxic to the brain.
    d. provide more fuel than fatty acids.

Short Answer Questions

1. The atoms needed to make protein are:

    a.   c.

    b.   d.

2. The three common parts of all amino acids are:

    a.   c.

    b.

3. Functions of proteins include:

    a.   f.

    b.   g.

    c.   h.

    d.   i.

    e.

4. The quality of dietary protein depends on:

    a.

    b.

5. The classic protein (-energy) deficiency diseases are:

    a.   b.

Problem Solving

1. Give the recommendations for protein (10-35 percent of total kcalories) in a 2000-kcalorie diet. How many kcalories should come from protein (provide range)?

2. How many grams of protein should contribute to the above recommendation?

3. If a meal provides a total of 570 kcalories, and 130 kcalories are from protein, what percentage of the kcalories is from protein?

4. What is the protein requirement for a person whose appropriate weight is 176 lb.?

5. A person's total kcalorie requirement is 2500. If this person consumes 950 kcalories from protein, what percent of kcalories are from protein?

6. How many grams of protein did this person consume in the above question?

7. Is this amount within the recommended range for protein intake?

8. If a person's protein requirement is 175 grams per day, how many kcalories should be consumed from protein?

9. If a person's protein requirement is 75 grams per day, what is the person's body weight in kg and in pounds?

10. If a person weighs 165 pounds, how many grams of protein should be consumed daily?

Figure Identification

Identify these structures and reaction.

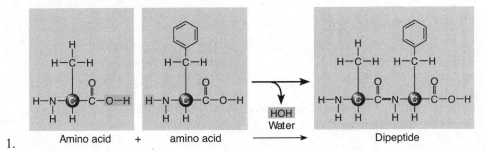

1.   Amino acid   +   amino acid   Dipeptide

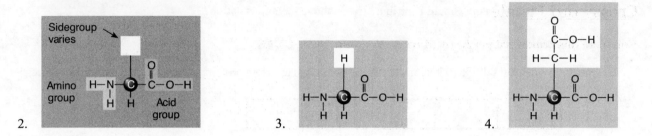

2.

3.

4.

Crossword Puzzle

Complete this crossword puzzle by Mary A. Wyandt, Ph.D., CHES.

Across:	Down:
1. large proteins of the blood and body fluids, produced by the immune system in response to the invasion of the body by foreign molecules	2. chemical messengers secreted by a variety of endocrine glands in response to altered conditions in the body
4. the protein material from which connective tissues and the foundation of bones and teeth are made	3. the basic substance that gives form to a developing structure; in the body, the formative cells from which teeth and bones grow
8. proteins that facilitate chemical reactions without being changed in the process; protein catalysts	5. the swelling of body tissue caused by excessive amounts of fluids in the interstitial spaces
9. the body's ability to recognize and eliminate foreign invaders	6. compounds composed of carbon, hydrogen, oxygen, and nitrogen atoms, arranged into amino acids linked in a chain
10. a globular protein in red blood cells that carries oxygen from the lungs to the cells throughout the body	7. substances that elicit the formation of antibodies or an inflammation reaction form the immune system

⑥ Chapter 6 Answer Key ⑥

Summing Up

1. amino
2. synthesized
3. Essential
4. proteins
5. energy
6. hemoglobin
7. proteomics
8. genes
9. genetic
10. transcription
11. translation
12. expression
13. enzymes
14. acid-base
15. Antibodies
16. clotting
17. pigments
18. amino
19. complete
20. proportions
21. Animal
22. protein
23. energy
24. malnutrition
25. problems
26. 0.8
27. amino acid

Chapter Study Questions

1. Like carbohydrates and fats, proteins contain carbon, hydrogen, and oxygen, but proteins also contain nitrogen. Amino acid structures have an amino group, an acid group, a hydrogen, and a side chain which makes each amino acid different from the others.

2. Amino acids are linked together to form proteins. The sequence and special characteristics of the side chains determine the protein's shape. The essential amino acids are histidine, isoleucine, leucine, lysine, methionine, phenylalanine, threonine, tryptophan, and valine.

3. In the mouth, chewing and crushing moisten protein-rich foods and mix them with saliva to be swallowed. In the stomach, stomach acid uncoils protein strands and activates enzymes; pepsin and HCl break protein down into smaller polypeptides. In the small intestine, pancreatic and small intestinal enzymes split polypeptides further into dipeptides, tripeptides, and amino acids, then enzymes on the surface of the small intestinal cells hydrolyze these peptides and the cells absorb them.

4. DNA in the nucleus of each cell serves as a template to make strands of messenger RNA. Each messenger RNA strand carries instructions for some protein the cell needs; the messenger RNA leaves the nucleus, and attaches itself to the protein-making machinery of the cell. Transfer RNA carry amino acids to the messenger RNA which dictates the sequence in which they will snap into place; this lines up the amino acids in sequence. The amino acids are then linked together in sequence, the completed protein strand is released, and later, the messenger RNA is degraded and the transfer RNA are re-used.

5. Proteins serve as enzymes, help maintain the body's fluid balance by attracting water, help maintain acid-base balance by acting as buffers, act against disease agents as antibodies, regulate body processes as hormones, transport nutrients and other molecules into and out of cells, help clot blood, help make scar tissue and bones, and serve as light-sensitive visual pigments.

6. Protein catalysts that facilitate the synthesis of larger compounds from smaller ones and hydrolysis of larger compounds to smaller ones without being affected in the process. Hormones are chemical messengers that are secreted by a variety of endocrine glands in response to altered conditions in the body.

7. The body uses amino acids for proteins or nonessential amino acids, for other compounds such as for synthesis of the neurotransmitters norepinephrine and epinephrine and melanin, or for energy. Deamination is removal of the amino group from a compound such as an amino acid. Nitrogen balance: the amount of nitrogen consumed compared with the amount of nitrogen excreted. Zero N balance: normal, healthy adult and lactating mothers; positive N balance: growing children and pregnant women; negative N balance: people who are sick or in trauma and people with kidney disease.

8. Its supply of a balance of the essential amino acids and its digestibility. A complete protein is a protein containing all the amino acids essential in human nutrition in amounts adequate for human use.

9. Vegetarians can obtain protein from legumes, nuts, vegetables, grains and (in some cases) eggs and milk products.

10. Protein-energy malnutrition, poor growth in children, and weight loss and wasting in adults. Maramus is the disease of starvation and kwashiorkor is the deficiency disease caused by inadequate protein in the presence of adequate food energy. The distinction is that someone with kwashiorkor has adequate energy, while someone with maramus has both inadequate protein and inadequate energy. Both conditions cause loss of body protein tissue.

11. Diets too high in protein offer no benefits. They often contribute fat-rich foods to diets of people who need to lose weight. In the small infant, the accumulation of amino acids stresses the kidneys and liver which have to metabolize and excrete the excess nitrogen; this may cause acidosis, dehydration, diarrhea, elevated blood ammonia, elevated blood urea, and fever. High-protein diets may promote calcium losses and deplete the bones of this mineral; if much animal-derived protein is eaten, this may contribute to the development of heart disease and cancer.

12. The quality of the protein eaten, kcalories and other nutrients consumed, a person's lean body mass, and a person's state of health.

13. No benefits, but risks include death, abnormal heart rhythms, severe amino acid imbalances and toxicities. Taking tryptophan supplements can cause eosinophilia myalgia syndrome (EMS).

Sample Test Questions

1. d (p. 181)	9. f (p. 192)	17. c (p. 189)	25. c (p. 197)
2. a (p. 183)	10. a (p. 192)	18. b (p. 191)	26. a (p. 199)
3. c (p. 181)	11. e (p. 192)	19. d (p. 192)	27. b (p. 201)
4. b (p. 190)	12. c (pp. 182-183)	20. b (p. 193)	28. c (p. 201)
5. a (p. 191)	13. b (pp. 195, 198)	21. a (p. 194)	29. d (p. 202)
6. b (p. 190)	14. d (p. 187)	22. f (p. 195)	30. c (p. 203)
7. c (p. 192)	15. a (p. 187)	23. b (p. 195)	
8. d (p. 190)	16. d (p. 187)	24. c (p. 196-197)	

Short Answer Questions

1. carbon; hydrogen; oxygen; nitrogen
2. amino group (NH_2); acid group (COOH); hydrogen (H)
3. enzymes; fluid balance; acid-base balance; antibodies; hormones; transport proteins; blood clotting; connective tissue; visual pigments
4. a balance of essential amino acids; protein digestibility
5. kwashiorkor; marasmus

Problem Solving

1. 2000 x .10 = 200 kcalories
   2000 x .35 = 700 kcalories
   200-700 kcalories
2. 200 divided by 4 = 50 grams
   700 divided by 4 = 175
   50 to 175 grams of protein
3. 130 kcal divided by 570 kcal = 23%
4. 176 lb. divided by 2.2 lb./kg = 80 kg
   0.8 g/kg x 80 kg = 64 g protein per day
5. 950 divided by 2500 = 38%
6. 950 divided by 4 = 238 grams protein
7. The recommended range is 10 to 35 percent. This is 38 percent, and therefore this amount exceeds the recommendation for protein intake.
8. 175 x 4 = 700 kcalories
9. 75 divided by 0.8 = 93.75 kg x 2.2 = 206 pounds
10. 165 divided by 2.2 = 75 kg x 0.8 = 60 grams of protein

Figure Identification

1. condensation of 2 amino acids to form a dipeptide
2. amino acid structure
3. glycine
4. aspartic acid

Crossword Puzzle

1. antibodies
2. hormones
3. matrix
4. collagen
5. edema
6. proteins
7. antigens
8. enzymes
9. immunity
10. hemoglobin

✆ Chapter 7 – Metabolism: ✆ Transformations and Interactions

Chapter Outline

I. Chemical Reactions in the Body
    A. The Site of Metabolic Reactions—Cells
    B. The Building Reactions—Anabolism
    C. The Breakdown Reactions—Catabolism
    D. The Transfer of Energy in Reactions—ATP
    E. The Helpers in Metabolic Reactions—
       Enzymes and Coenzymes

II. Breaking Down Nutrients for Energy
    A. Glucose
       1. Glucose-to-Pyruvate
       2. Pyruvate's Options
       3. Pyruvate-to-Lactate
       4. Pyruvate-to-Acetyl CoA
       5. Acetyl CoA's Options
    B. Glycerol and Fatty Acids
       1. Glycerol-to-Pyruvate
       2. Fatty Acids-to-Acetyl CoA
       3. Fatty Acids Cannot Be Used to
          Synthesize Glucose
    C. Amino Acids
       1. Amino Acids-to-Acetyl CoA
       2. Amino Acids-to-Glucose
       3. Deamination
       4. Transamination
       5. Ammonia-to-Urea in the Liver
       6. Urea Excretion via the Kidneys
    D. Breaking Down Nutrients for Energy—In
       Summary

E. The Final Steps of Catabolism
    1. The TCA Cycle
    2. The Electron Transport Chain
    3. The kCalories-per-Gram Secret
       Revealed

III. Energy Balance
    A. Feasting—Excess Energy
       1. Excess Protein
       2. Excess Carbohydrate
       3. Excess Fat
    B. The Transition from Feasting to Fasting
    C. Fasting—Inadequate Energy
       1. Glucose Needed for the Brain
       2. Protein Meets Glucose Needs
       3. The Shift to Ketosis
       4. Suppression of Appetite
       5. Slowing of Metabolism
       6. Symptoms of Starvation

IV. Alcohol and Nutrition
    A. Alcohol in Beverages
    B. Alcohol in the Body
    C. Alcohol Arrives in the Liver
    D. Alcohol Disrupts the Liver
    E. Alcohol Arrives in the Brain
    F. Alcohol and Malnutrition
    G. Alcohol's Short-Term Effects
    H. Alcohol's Long-Term Effects
    I. Personal Strategies

Summing Up

The principal compounds derived from carbohydrate, (1)_____, and protein in the diet are (2)_____, glycerol, fatty acids, and amino acids. Glucose may be anabolized to (3)_____ or catabolized to (4)_____, which in turn yields acetyl CoA. Glycerol and fatty acids may be anabolized to (5)_____ or catabolized—glycerol to pyruvate, fatty acids to acetyl CoA. Amino acids may be anabolized to (6)_____ or catabolized (after deamination) to pyruvate, acetyl CoA, or intermediates that enter the (7)_____ cycle directly. When amino acids are deaminated, the nitrogen removed is combined with carbon dioxide to form urea and excreted.

Pyruvate is convertible to glucose, but (8)_____ CoA is not. Hence fatty acids, which break down to multiple units of acetyl CoA, cannot serve to generate (9)_____ for the body. All three energy-yielding nutrients are convertible to acetyl CoA, however, and hence can be used to manufacture body (10)_____, or provide (11)_____.

If a person (12)_____ or if carbohydrate is undersupplied, lean body tissue is catabolized to meet the brain's need for (13)_____. Within a day or so after the start of a fast, a person is in (14)_____. (15)_____ acids are metabolized to ketone bodies, which can meet some of the (16)_____ energy need. Weight loss may be dramatic, but (17)_____ loss may be slower than on a moderate, balanced low-kcalorie diet. To design such a moderate diet requires adjusting energy balance so that the intake is (18)_____, the output is (19)_____, or both.

Chapter Study Questions

1. Define metabolism, anabolism, and catabolism; give an example of each.

2. Name one of the body's high-energy molecules, and describe how is it used.

3. What are coenzymes, and what service do they provide in metabolism?

4. Name the four basic units, derived from foods, used by the body in metabolic transformations. How many carbons are in the "backbones" of each?

5. Define aerobic and anaerobic metabolism. How does insufficient oxygen influence metabolism?

6. How does the body dispose of excess nitrogen?

7. Summarize the main steps in the metabolism of glucose, glycerol, fatty acids, and amino acids.

8. Describe how a surplus of the three energy nutrients contributes to body fat stores.

9. What adaptations does the body make during a fast? What are ketone bodies? Define ketosis.

10. Distinguish between a loss of fat and a loss of weight, and describe how both might happen.

Chapter Glossary

- **acetyl CoA:** a 2-carbon compound (acetate, or acetic acid) to which a molecule of CoA is attached.
- **aerobic:** requiring oxygen.
- **ammonia:** a compound with the chemical formula NH_3; produced during the deamination of amino acids.
- **anabolism:** reactions in which small molecules are put together to build larger ones. Anabolic reactions require energy.
- **anaerobic:** not requiring oxygen.
- **ATP** or **adenosine triphosphate:** a common high-energy compound composed of a purine (adenine), a sugar (ribose), and three phosphate groups.
- **catabolism:** reactions in which large molecules are broken down to smaller ones. Catabolic reactions release energy.
- **CoA:** coenzyme A; the coenzyme derived from the B vitamin pantothenic acid and central to energy metabolism.
- **coenzymes:** complex organic molecules that work with enzymes to facilitate the enzymes' activity. Many coenzymes have B vitamins as part of their structures.
- **Cori cycle:** the path from muscle glycogen to glucose to pyruvate to lactate (which travels to the liver) to glucose (which can travel back to the muscle) to glycogen; named after the scientist who elucidated this pathway.
- **coupled reactions:** pairs of chemical reactions in which some of the energy released from the breakdown of one compound is used to create a bond in the formation of another compound.
- **electron transport chain:** the final pathway in energy metabolism that transports electrons from hydrogen to oxygen and captures the energy released in the bonds of ATP.
- **energy metabolism:** all the reactions by which the body obtains and expends the energy from food.
- **fatty acid oxidation:** the metabolic breakdown of fatty acids to acetyl CoA; also called beta oxidation.
- **fuel:** compounds that cells can use for energy. The major fuels include glucose, fatty acids, and amino acids; other fuels include ketone bodies, lactate, glycerol, and alcohol.
- **glycolysis:** the metabolic breakdown of glucose to pyruvate. Glycolysis does not require oxygen (anaerobic).
- **keto acid:** an organic acid that contains a carbonyl group (C=O).
- **lactate:** a 3-carbon compound produced from pyruvate during anaerobic metabolism.
- **metabolism:** the sum total of all the chemical reactions that go on in living cells.
- **mitochondria:** the cellular organelles responsible for producing ATP aerobically; made of membranes (lipid and protein) with enzymes mounted on them.
- **oxaloacetate:** a carbohydrate intermediate of the TCA cycle.

- **photosynthesis:** the process by which green plants use the sun's energy to make carbohydrates from carbon dioxide and water.
- **pyruvate:** a 3-carbon compound that plays a key role in energy metabolism.
- **TCA cycle** or **tricarboxylic acid cycle:** a series of metabolic reactions that break down molecules of acetyl CoA to carbon dioxide and hydrogen atoms; also called the Kreb's cycle after the biochemist who elucidated its reactions.
- **transamination:** the transfer of an amino group from one amino acid to a keto acid, producing a new nonessential amino acid and a new keto acid.
- **urea:** the principal nitrogen-excretion product of protein metabolism. Two ammonia fragments are combined with carbon dioxide to form urea.

Sample Test Questions

Select the best answer for each question.

1. Photosynthesis is:
   a. the sum total of chemical reactions.
   b. a reaction that releases energy.
   c. fuel for the body.
   d. the process of plants making simple sugars.

2. Compounds that the body can use for energy are called:
   a. fuel.   c. minerals.
   b. vitamins.   d. enzymes.

3. Energy, water and carbon dioxide are released during what process?
   a. energy release   c. anabolism
   b. metabolism   d. energy storage

4. Nutrients which are oxidized in the body to yield energy are:
   a. water, carbohydrate, fat, protein, vitamins, and minerals.
   b. carbohydrate, fat, protein, vitamins, and minerals.
   c. carbohydrate, fat, protein, and vitamins.
   d. carbohydrate, fat, and protein.
   e. carbohydrate and vitamins.

5. Reactions in which compounds are broken down into simple molecules are called _____ reactions.
   a. anabolic   c. gluconeogenic
   b. catabolic   d. erogenic

6. Small organic molecules that work to facilitate the activity of enzymes are:
   a. adenosines.   c. coenzymes.
   b. phosphates.   d. acetyl CoA.

7. Which of the following statements is **false** regarding coupled reactions?
   a. Coupled reactions do not include the hydrolysis of ATP.
   b. Coupled reactions include the use of ATP during catabolism to power an anabolic reaction.
   c. Coupled reactions involve a transfer of energy.
   d. Coupled reactions have been referred to as a metabolic duet.

8. A three-carbon compound reversibly convertible to glucose is:
   a. an amino acid.   d. a fatty acid.
   b. pyruvate.   e. glucose.
   c. acetyl CoA.

9. When glucose is split in half, it yields two molecules of:
   a. acetate.
   b. glycerol.
   c. pyruvate.
   d. water.
   e. monosaccharides.

10. The part(s) of triglycerides that can be made into glucose is/are:
    a. short-chain fatty acids.
    b. medium-chain fatty acids.
    c. long-chain fatty acids.
    d. all fatty acids.
    e. glycerol.

11. The process of glucose being split is called:
    a. the electron transport chain.
    b. the TCA cycle.
    c. gluconeogenesis.
    d. glycolysis.

12. Pyruvate is converted to lactate in:
    a. an anaerobic pathway.
    b. an aerobic pathway.
    c. pyruvate oxidation.
    d. glycolysis.

13. The concentration of lactate increases dramatically after:
    a. aerobic exercise.
    b. high-intensity exercise.
    c. low-intensity exercise.
    d. rest.

14. If the cells need energy and oxygen is available:
    a. lactate builds up.
    b. the Cori cycle is activated.
    c. pyruvate molecules enter the mitochondria.
    d. anaerobic metabolism begins.

15. When energy nutrients are metabolized to yield energy, this is called:
    a. oxidation.
    b. pathways.
    c. anaerobic reactions.
    d. coenzymes.

16. The reaction which removes the NH_2 group from an amino acid is called:
    a. transamination.
    b. deamination.
    c. nitrogenation.
    d. hydrogenation.

17. Nitrogen from amino groups is excreted from the body in:
    a. urea.
    b. ammonia.
    c. nitrogen gas.
    d. amino acids.

18. Which of the following is **false** about the TCA cycle?
    a. It regenerates acetyl CoA.
    b. It requires oxaloacetate in the first step and synthesizes it in the last step.
    c. It requires ample carbohydrate.
    d. Each turn of the TCA cycle releases 8 electrons.

19. Which of the following is **false** about the electron transport chain?
    a. It is the final pathway.
    b. It results in release of ATP to the cytoplasm.
    c. It consists of proteins that serve as carriers.
    d. The carriers are on the inner membrane of the DNA.

20. When glucose consumption is in excess of body needs, the excess glucose is:
    a. not absorbed from the small intestine.
    b. excreted in the feces.
    c. stored as glucose.
    d. stored as glycogen only.
    e. stored as glycogen and fat.

21. Energy is stored in the body for future use as:
    a. triglycerides.
    b. glycerol.
    c. cholesterol.
    d. lecithin.

22. Which of the following statements is **true** regarding feasting?
    a. The storage form depends on the type of energy-yielding nutrient consumed.
    b. Food energy excess cannot be stored in the body.
    c. Excess carbohydrate and fat are the only nutrients that can make you fat.
    d. Excess protein can be stored as fat.

23. Which of the following statements is **false** regarding fasting?
    a. Glucose is the preferred fuel for cells of the brain and nervous system.
    b. During fasting, body protein stores break down.
    c. During fasting, ketone bodies are produced.
    d. During fasting, fat cannot be used for energy.

24. During the second phase of fasting the body adapts by producing an alternative energy source:
    a. glycogen.
    b. ammonia.
    c. pyruvate.
    d. ketone bodies.

25. Elevated urine ketones are called:
    a. acids.
    b. ketosis.
    c. ketonuria.
    d. acidosis.

26. Which of the following is **false** regarding ketosis?
    a. It induces a loss of appetite.
    b. It causes acetone breath.
    c. It increases metabolism.
    d. It causes the pH of the blood to drop.

27. Short-term fasting usually produces:
    a. weight loss.
    b. fat loss.
    c. more lean tissue.
    d. all of the above

28. Symptoms of starvation include all but which of the following?
    a. wasting
    b. raised metabolism
    c. lowered body temperature
    d. reduced resistance to disease

29. Psychological effects of food deprivation include:
    a. muscle wasting, slowed heart rate, and depression.
    b. significant loss of body fat, loss of water, and anxiety.
    c. significant loss of body weight, hunger, and elation.
    d. depression, anxiety, and food-related dreams.

30. Which of the following statements is **false** regarding starvation?
    a. It slows energy output.
    b. Its progression differs among a starving child, an adult fasting for religious reasons, and the person with anorexia.
    c. It produces physical and emotional hazards.
    d. It increases the breakdown of fat.

Short Answer Questions

1. The energy nutrients are broken down into these basic units:

    a. carbohydrate: _____

    b. lipids: _____

    c. lipids: _____

    d. protein: _____

2. How many carbons is in each of these compounds?

    a. glucose: _____

    b. glycerol: _____

    c. fatty acids: _____

    d. amino acids: _____

    e. pyruvate: _____

    f. acetyl CoA: _____

Problem Solving

1. To lose 3 pounds, a person needs to adjust the energy budget so that intake is less than output by how many kcalories?

2. What should the daily kcalorie deficit be if a person wanted to lose 3 pounds in 30 days?

3. If the person were to consume 750 kcalories less per day than the energy output, how long would it take to lose 3 pounds?

Figure Identification

Identify the following structures, pathways, or processes.

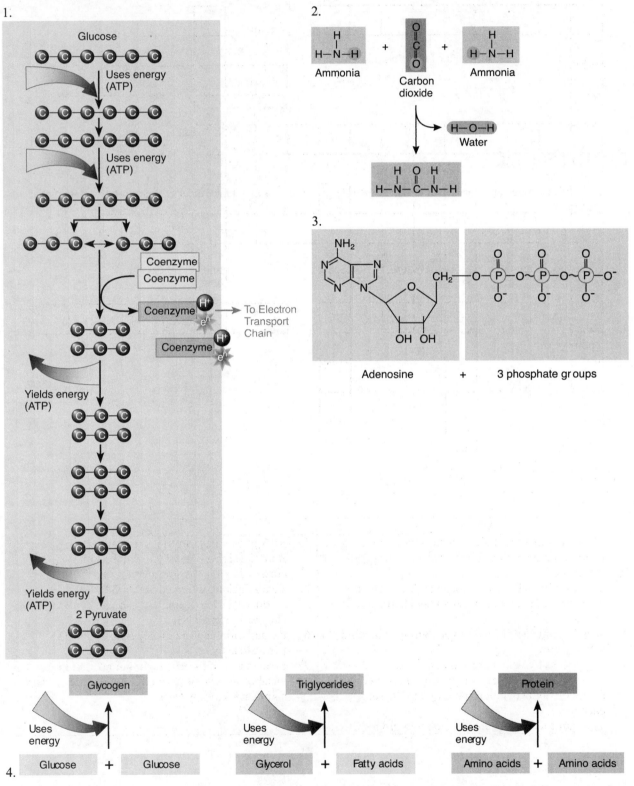

1.

Glucose

Uses energy (ATP)

Uses energy (ATP)

Coenzyme

Coenzyme

Coenzyme H⁺ e⁻ → To Electron Transport Chain

Coenzyme H⁺ e⁻

Yields energy (ATP)

Yields energy (ATP)

2 Pyruvate

2.

Ammonia + Carbon dioxide + Ammonia

→ H—O—H Water

3.

Adenosine + 3 phosphate groups

4.

Glycogen

Uses energy

Glucose + Glucose

Triglycerides

Uses energy

Glycerol + Fatty acids

Protein

Uses energy

Amino acids + Amino acids

Crossword Puzzle

Complete this crossword puzzle by Mary A. Wyandt, Ph.D., CHES.

Across:	Down:
1. substances that facilitate enzyme action; this includes both organic coenzymes and inorganic substances	1. pairs of chemical reactions in which energy released from the breakdown of one compound is used to create a bond in the formulation of another
3. a series of metabolic reactions that break down molecules of acetyl CoA to CO_2 and hydrogen atoms	2. a carbohydrate intermediate of the TCA cycle
4. an acid produced from pyruvate during anaerobic metabolism	5. produced during deamination, a compound with the chemical formula NH_3
8. the process by which green plants make carbohydrates from carbon dioxide and water using the green pigment chlorophyll to trap the sun's energy	6. the path from muscle glycogen to glucose to pyruvate to lactic acid
9. an organic acid that contains carbonyl group (C=O)	7. pyruvic acid, a 3-carbon compound, that in metabolism, can be derived form glucose, certain amino acids, or glycerol

⑥ Chapter 7 Answer Key ⑨

Summing Up

1. fat;
2. glucose
3. glycogen
4. pyruvate
5. triglycerides
6. protein
7. TCA
8. acetyl
9. glucose
10. fat
11. energy
12 fasts
13. glucose
14. ketosis
15. Fatty
16. brain's
17. fat
18. reduced
19. increased

Chapter Study Questions

1. Metabolism includes all chemical reactions, both anabolic and catabolic, that occur in living cells; for example, fructose is metabolized to glucose. Anabolic reactions put small molecules together to build larger ones; for example, glycerol + fatty acids = triglyceride. Catabolic reactions break down large molecules to smaller ones; for example, triglyceride = glycerol + fatty acids.
2. ATP is used as energy currency in metabolic reactions.
3. Coenzymes are small organic molecules that work with enzymes to facilitate the enzymes' activity; they enable enzymes to do their work.
4. Glucose, glycerol, fatty acids, and amino acids. Glucose has 6 carbons, glycerol has 3 carbons, fatty acids have multiples of 2 carbons, and amino acids have 2, 3, or more carbons.
5. Aerobic metabolism refers to chemical reactions that require oxygen; anaerobic reactions do not require oxygen. Insufficient oxygen favors anaerobic metabolism that results in lactic acid production.
6. When amino nitrogen is stripped from amino acids, ammonia is produced. The liver detoxifies ammonia before releasing it into the bloodstream by combining it with another waste product, carbon dioxide, to produce urea.
7. Glucose to pyruvate to acetyl CoA to carbon dioxide. Glycerol to pyruvate either to glucose or to acetyl CoA to carbon dioxide. Fatty acids either to pyruvate or to acetyl CoA to carbon dioxide. Protein can be broken down to amino acids and if used for energy, it must be deaminated (its nitrogen containing amino group must be removed). Some amino acids can be used to make pyruvate and others can be used to make acetyl CoA.
8. Each can be broken down to pyruvate and acetyl CoA which can be built up into fat (fat can also be stored directly).
9. The body draws on carbohydrate and fat reserves to spare protein. When reserves are exhausted, body protein is used to make glucose; as the fast continues, ketones are made to meet energy needs. The body reduces its energy output and fat loss. Ketone bodies are compounds produced during the incomplete breakdown of fat when glucose is not available. Ketosis is the buildup of ketone bodies in the blood and urine.
10. Weight loss on a low-carbohydrate diet is a loss of glycogen, protein, water, and minerals, not just a loss of fat. A moderate, balanced, low-kilocalorie diet is more likely to promote loss of fat.

Sample Test Questions

1. d (p. 213)
2. a (p. 213)
3. b (p. 213)
4. d (p. 213)
5. b (p. 216)
6. c (p. 216)
7. a (p. 216)
8. b (p. 218)
9. c (p. 219-220)
10. e (p. 222)
11. d (p. 219)
12. a (p. 220-221)
13. b (p. 221)
14. c (p. 221)
15. a (p. 223)
16. b (p. 225)
17. a (p. 225-226)
18. a (p. 227-228)
19. d (p. 229)
20. e (p. 233)
21. a (p. 232)
22. d (p. 232)
23. d (p. 234-235)
24. d (p. 235)
25. c (p. 235)
26. c (p. 235)
27. a (p. 235)
28. b (p. 236)
29. d (p. 236)
30. b (p. 235-236)

Short Answer Questions

1. a. glucose; b. & c. glycerol, fatty acids; d. amino acids
2. a. 6; b. 3; c. multiples of 2; d. 2 or 3 or more; e. 3; f. 2

Problem Solving

1. 3 lb x 3,500 kcal/lb. = 10,500 kcal
2. 10,500 kcal divided by 30 days = 350 kcal/day
3. 10,500 kcal divided by 750 kcal/day = 14 days

Figure Identification

1. glycolysis
2. urea synthesis
3. ATP
4. anabolic reactions

Crossword Puzzle

1A. cofactors
1D. coupled reactions
2. oxaloacetate
3. TCA cycle
4. lactic acid
5. ammonia
6. Cori cycle
7. pyruvate
8. photosynthesis
9. keto acid

✿ Chapter 8 – Energy Balance and Body Composition ✿

Chapter Outline

I. Energy Balance
II. Energy In: The kCalories Foods Provide
  A. Food Composition
  B. Food Intake
    1. Hunger
    2. Satiation
    3. Satiety
    4. Overriding Hunger and Satiety
    5. Sustaining Satiation and Satiety
    6. Message Central—The Hypothalamus
III. Energy Out: The kCalories the Body Expends
  A. Components of Energy Expenditure
    1. Basal Metabolism
    2. Physical Activity
    3. Thermic Effect of Food
    4. Adaptive Thermogenesis
  B. Estimating Energy Requirements
IV. Body Weight, Body Composition, and Health
  A. Defining Healthy Body Weight
    1. The Criterion of Fashion
    2. The Criterion of Health
    3. Body Mass Index
  B. Body Fat and Its Distribution
    1. Some People Need Less Body Fat
    2. Some People Need More Body Fat
    3. Fat Distribution
    4. Waist Circumference
    5. Other Measures of Body Composition

C. Health Risks Associated with Body Weight and Body Fat
    1. Health Risks of Underweight
    2. Health Risks of Overweight
    3. Cardiovascular Disease
    4. Diabetes
    5. Inflammation and the Metabolic Syndrome
    6. Cancer
    7. Fit and Fat versus Sedentary and Slim
V. Eating Disorders
  A. The Female Athlete Triad
    1. Disordered Eating
    2. Amenorrhea
    3. Osteoporosis
  B. Other Dangerous Practices of Athletes
  C. Preventing Eating Disorders in Athletes
  D. Anorexia Nervosa
    1. Characteristics of Anorexia Nervosa
      a. Self-Starvation
      b. Physical Consequences
    2. Treatment of Anorexia Nervosa
  E. Bulimia Nervosa
    1. Characteristics of Bulimia Nervosa
      a. Binge Eating
      b. Purging
    2. Treatment of Bulimia Nervosa
  F. Binge-Eating Disorder
  G. Eating Disorders in Society

Summing Up

Energy intake can be computed by adding up the (1)_____ values of the foods consumed or can be estimated by using (2)_____ system values. The energy available from food was originally determined by measuring the (3)_____ lost when the food was completely burned, with adjustments to account for the incomplete breakdown of food in the (4)_____. Machines designed to measure energy expenditure work by measuring heat lost from the body or (5)_____ and carbon dioxide exchanged. Data collected in this way have yielded tables from which to estimate (6)_____ needs.

The body's total energy output falls into two major categories: energy to support the (7)_____ or resting metabolism, and energy for (8)_____ activity. The basal metabolic rate is influenced primarily by the amount of (9)_____ body mass; it is also affected by fever, fasting, and (10)_____ secretions (especially epinephrine and thyroxin).

An energy deficit of (11)_____ kcalories is necessary for the loss of a pound of body fat. Rapid declines in body weight include loss of some fat, large amounts of (12)_____, and some (13)_____ tissues such as muscle proteins and bone minerals. Even over the long term, the composition of weight lost is normally (14)_____ percent fat and (15)_____ percent lean.

People of the same sex, age, and height may differ in (16)_____ not only due to differing densities of their bones and muscles. The weight compatible with good health depends on the individual. Women may have larger amounts of total body fat because of great quantities of (17)_____ fat. Health problems typically develop when body fat exceeds (18)_____ percent in young men and (19)_____ percent in young women.

The distribution of fat may be more critical than total (20)_____ of fat alone. (21)_____ fat is stored around the organs of the abdomen and is referred to as (22)_____ obesity. This obesity is associated with increased risks of heart disease, stroke, diabetes, hypertension, gallstones, and some forms of (23)_____. A person's (24)_____ circumference is the most practical indicator of fat distribution and central obesity.

Obesity entails a host of health hazards, in addition to (25)_____ and economic disadvantages. By contrast, (26)_____ is associated with an increased risk from wasting diseases, osteoporosis and medical stresses. Underweight people may be at risk for inability to preserve lean tissue during a fight against (27)_____ or a digestive disorder. (28)_____ may be a problem for underweight women.

Health risks for overweight include diabetes, hypertension, cardiovascular disease, (29)_____, some cancers, gallbladder disease, kidney stones, respiratory problems, and complications during pregnancy or (30)_____. Chronic inflammation accompanies obesity and contributes to (31)_____ diseases.

Chapter Study Questions

1. What are the consequences of an unbalanced energy budget?

2. Define hunger, appetite, satiation, and satiety and describe how each influences food intake.

3. Describe each component of energy expenditure. What factors influence each? How can energy expenditure be estimated?

4. Distinguish between body weight and body composition. What assessment techniques are used to measure each?

5. What problems are involved in defining "ideal" body weight?

6. What is central obesity and what is its relationship to disease?

7. What risks are associated with excess body weight and excess body fat?

Chapter Glossary

- **adaptive thermogenesis:** adjustments in energy expenditure related to changes in environment such as cold and to physiological events such as overfeeding, trauma, and changes in hormone status.
- **air displacement plethysmography:** estimates body composition by having a person sit inside a chamber while computerized sensors determine the amount of air displaced by the person's body.
- **appetite:** the integrated response to the sight, smell, thought, or taste of food that initiates or delays eating.
- **basal metabolic rate (BMR):** the rate of energy use for metabolism under specified conditions: after a 12-hour fast and restful sleep, without any physical activity or emotional excitement, and in a comfortable setting. It is usually expressed as kcalories per kilogram body weight per hour.
- **basal metabolism:** the energy needed to maintain life when a body is at complete digestive, physical, and emotional rest.
- **bioelectrical impedance:** a method for estimating body fat using low-intensity electrical current.
- **body composition:** the proportions of muscle, bone, fat, and other tissue that make up a person's total body weight.
- **body mass index (BMI):** an index of a person's weight in relation to height, determined by dividing the weight (in kilograms) by the square of the height (in meters).
- **bomb calorimeter:** an instrument that measures the heat energy released when foods are burned, thus providing an estimate of the potential energy of the foods.
- **central obesity:** excess fat around the trunk of the body; also called *abdominal fat* or *upper body fat*.
- **direct calorimetry:** the measurement of energy output as heat energy.
- **dual energy X-ray absorptiometry (DEXA):** uses two low-dose X-rays that differentiate among fat-free soft tissue (lean body mass), fat tissue, and bone tissue, providing a precise measurement of total fat and its distribution in all but extremely obese subjects.
- **fatfold measures:** estimate body fat by using a caliper to gauge the thickness of a fold of skin on the back of the arm (over the triceps), below the shoulder blade (subscapular), and in other places (including lower-body sites) and then comparing these measurements with standards.
- **hunger:** the painful sensation caused by a lack of food that initiates food-seeking behavior.

- **hydrodensitometry:** measures body density by weighing the person first on land and then again while submerged in water.
- **hypothalamus**: a brain center that controls activities such as maintenance of water balance, regulation of body temperature, and control of appetite.
- **indirect calorimetry:** the estimation of energy output from measures of the amount of oxygen used and carbon dioxide eliminated.
- **insulin resistance**: the condition in which a normal amount of insulin produces a subnormal effect, resulting in an elevated fasting glucose; a metabolic consequence of obesity that precedes type 2 diabetes.
- **intra-abdominal fat:** fat stored within the abdominal cavity in association with the internal abdominal organs, as opposed to the fat stored directly under the skin (subcutaneous fat).
- **lean body mass**: the weight of the body minus the fat content.
- **neuropeptide Y**: a chemical produced in the brain that stimulates appetite, diminishes energy expenditure, and increases fat storage.
- **overweight:** body weight above some standard of acceptable weight that is usually defined in relation to height (such as BMI).
- **physiological fuel value:** the number of kcalories that the human body derives from a food, as contrasted with the number of kcalories determined by calorimetry.
- **resting metabolic rate (RMR):** similar to the basal metabolic rate (BMR), a measure of the energy use of a person at rest in a comfortable setting, but with less stringent criteria for recent food intake and physical activity. Consequently, the RMR is slightly higher than the BMR.
- **satiating**: having the power to suppress hunger and inhibit eating.
- **satiation**: the feeling of satisfaction that occurs during a meal and halts eating.
- **satiety:** the feeling of fullness and satisfaction that occurs after a meal and inhibits eating until the next meal.
- **stress eating:** eating in response to arousal.
- **thermic effect of food (TEF):** an estimation of the energy required to process food (digest, absorb, transport, metabolize, and store ingested nutrients); also called the *specific dynamic effect (SDE) of food*, or the *specific dynamic activity (SDA) of food*. The sum of the TEF and any increase in the metabolic rate due to overeating is known as *diet-induced thermogenesis (DIT)*.
- **thermogenesis:** the generation of heat; used in physiology and nutrition studies as an index of how much energy the body is expending.
- **underweight:** body weight below some standard of acceptable weight that is usually defined in relation to height (such as BMI).
- **waist circumference**: an anthropometric measurement used to assess a persons abdominal fat.

Sample Test Questions

Select the best answer for each question.

1. For each _____ kcal eaten in excess, one pound of body fat is stored.
   a. 2,000
   b. 2,500
   c. 3,000
   d. 3,500

2. An instrument that measures the heat energy released when foods are burned is a:
   a. body mass index.
   b. basal metabolic rate.
   c. bomb calorimeter.
   d. heat food release mechanism.

3. The feeling of fullness that occurs during a meal and halts eating is called:
   a. satiation.
   b. satiety.
   c. purging.
   d. appetite.

4. If something has the power to suppress hunger and inhibit eating, you describe it as:
   a. appetite control.
   b. effective binging and purging.
   c. satiating.
   d. satiation.

5. Which of the following statements is **true** about resting metabolic rate?
    a. It is slightly lower than the BMR.
    b. It is lower than the thermic effect of food.
    c. It is the same as diet-induced thermogenesis.
    d. It is a measure of energy use of a person at rest in a comfortable position.

6. Which of the following describes the process of thermogenesis?
    a. burning of fat
    b. synthesis of fat
    c. generation of heat
    d. generation of water

7. What method is used to measure the amount of heat given off by the body?
    a. bomb calorimetry
    b. basal calorimetry
    c. direct calorimetry
    d. indirect calorimetry

8. Additional energy that is spent when a person must adapt to extremely cold conditions is called:
    a. basal metabolic rate.
    b. resting metabolism.
    c. adaptive thermogenesis.
    d. rebound thermogenesis.

9. The number of kcalories that the body derives from a food is called:
    a. the physiological fuel value.
    b. the calorimetry value.
    c. less precise than direct calorimetry.
    d. less precise than computing values based on grams of energy-yielding nutrients.

10. The amount of energy a food contains may be measured by burning the food and measuring the:
    a. vitamins produced.
    b. heat released.
    c. water produced.
    d. water released.
    e. weight of the ash remaining.

11. The energy used by the body completely at rest with no food recently eaten is:
    a. specific dynamic energy.
    b. basal metabolic energy.
    c. active energy.
    d. total kcalories expended.
    e. zero energy.

12. Which of the following will reduce BMR?
    a. fever
    b. fasting
    c. a cold environment
    d. high thyroid activity

13. A measure of energy output that is less precise than the BMR is:
    a. physical activity.
    b. thermic effect.
    c. specific dynamic effect.
    d. resting metabolic rate.

14. What is the major factor that determines metabolic rate?
    a. age
    b. gender
    c. amount of fat tissue
    d. amount of lean body mass

15. Which of the following increases the metabolic rate?
    a. fever
    b. fasting
    c. inactivity
    d. malnutrition

16. Which of the following statements is **true** regarding satiation?
    a. It is achieved most by high carbohydrate intake.
    b. Fat has a weak effect on satiation.
    c. Protein is the most satiating.
    d. Satiation is not influenced by energy-yielding nutrients.

17. Which chemical causes carbohydrate cravings?
    a. proteins   c. estrogen
    b. neuropeptide Y   d. enzymes

18. What would be the approximate weight gain of a person who consumes an excess of 500 kcal daily for one month?
    a. 0.5 lb.   c. 3 lbs.
    b. 2 lbs.   d. 4 lbs.

19. A BMI of 27 is considered:
    a. underweight.   c. overweight.
    b. normal.   d. obese.

20. If you eat 2,400 kcalories a day, about _____ will be used for the thermic effect of food.
    a. 24 kcalories   d. 10 kcalories
    b. 240 kcalories   e. 3,500 kcalories
    c. 100 kcalories

21. Degree of fatness can most accurately be assessed by:
    a. the skinfold test.
    b. the accepted style of the culture.
    c. undressing and standing before a mirror to see how you look to yourself.
    d. the fit of your clothes.
    e. comparing weight with standard weight charts.

22. Which of the following statements is true?
    a. Fashion is helpful when assessing ideal body weights.
    b. Fashion has contributed to establishing healthy standards for body weight.
    c. Perceived body size is almost identical to actual body size.
    d. Fashion presents unrealistic and unhealthy ideals for body size.

23. An index of a person's weight in relation to height is called:
    a. body mass index.   c. ideal body weight index.
    b. height-to-weight index.   d. desirable body weight index.

24. Which of the following statements is **false** about the body mass index?
    a. It reflects height and weight measures.
    b. It does not reflect body composition.
    c. It is ideal for estimating health risk.
    d. It may misclassify muscular athletes as overweight or obese.

25. Which measures of waist circumferences for women and men indicate health risk?
    a. 22 or more inches for women and 25 or more inches for men
    b. 25 or more inches for women and 30 or more inches for men
    c. 30 or more inches for women and 35 or more inches for men
    d. 35 or more inches for women and 40 or more inches for men

26. Which statement is true regarding health risks?
    a. They indicate correlations with disease.
    b. They reveal causes of disease.
    c. BMI is not a health risk.
    d. They provide information about life expectancy with 99 percent accuracy.

27. Underweight people may face the following risks:
    a. malnutrition.
    b. infertility.
    c. trouble during wasting diseases.
    d. all of the above

28. Overweight persons are faced with the possibility of these effects of obesity:
    1. earlier death due to a host of physical problems.
    2. precipitation of diabetes.
    3. lowered accident rate due to the protection from adipose tissue.
    4. increased risk of cardiovascular disease.
    5. greater economic success because they are left alone to concentrate on their jobs.

    a. 1, 3, 5
    b. 1, 2, 4
    c. 1, 2, 5
    d. 2, 3, 4
    e. 3, 4, 5

29. Which of the following is considered as strong a risk factor for heart attack as high blood cholesterol, hypertension and smoking?
    a. overweight
    b. obesity
    c. central obesity
    d. cancer

30. Accumulation of fat and activation of genes that code for proteins involved in inflammation causes:
    a. obesity.
    b. diabetes.
    c. metabolic syndrome.
    d. insulin resistance.

Short Answer Questions

1. The two major contributors to energy output are:

   a.   b.

Problem Solving

1. Estimate the energy required for basal metabolism for a 175-pound man for a 24-hour period.

2. A person's EER is 2500 kcalories per day. This person wants to lose 1 pound of fat a week. How many kcalories should they consume each day?

3. A person's body weight is 150 pounds and this person is 75 inches tall. Calculate the BMI.

4. A person's body weight is 132 pounds and this person is 5 feet 6 inches tall. Calculate the BMI.

5. A person's EER is 3000 kcalories per day. If 65% is BMR, 10% is thermic effect of food, and 25% is physical activities, how many kcalories is this person expending in each component of energy expenditure?

6. A person has a BMI of 26 and is 5'8". She desires to have a BMI of 24. Calculate her desired body weight.

7. Calculate how much energy a 150-pound person would expend doing 30 minutes of vigorous aerobic dance.

8. Calculate how much energy a 170-pound person would expend doing 20 minutes of running at 9 mph.

9. Compute the daily energy needs for a women, age 47, who is 5 feet 6 inches tall, weighs 130 lbs., and is lightly active.

10. Compute the daily energy needs for a man, age 21, who is 5 feet 10 inches tall, weighs 142 lbs., and is active.

Crossword Puzzle

Complete this crossword puzzle by Mary A. Wyandt, Ph.D., CHES.

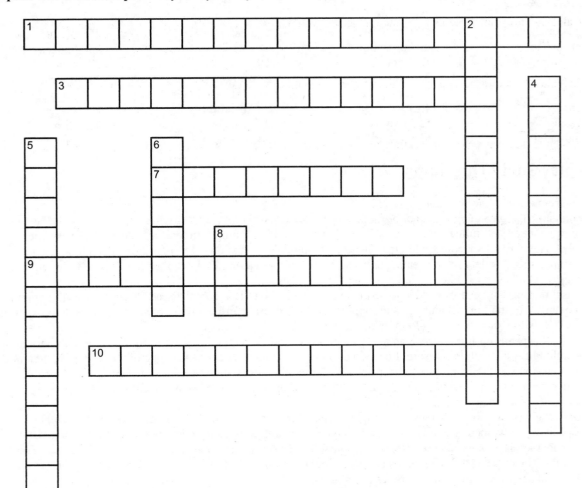

Across:		Down:	
1.	a method of measuring body density; the person is first weighed and then submerged in water	2.	a generation of heat; used in physiology and nutrition studies as an index of how much energy the body is spending
3.	a clinical estimate of total body fatness in which the thickness of a fold of skin is measured with a caliper	4.	a brain center that controls activities such as maintenance of water balance, regulation of body temperature, and control of appetite
7.	a type of calorimetry measurement that estimates the energy output from measures of the amount of oxygen used and carbon dioxide eliminated	5.	the weight of the body minus the fat content
		6.	a type of calorimetry measurement that measures energy output as heat energy
9.	the proportions of muscle, bone, fat, and other tissue that make up a person's total body weight	8.	abbreviation for an index of a person's weight in relation to height, determined by dividing the weight (kg) by the square of the height (m)
10.	the energy needed to maintain life when a body is at complete rest after a 12-hour fast		

◌ Chapter 8 Answer Key ◌

Summing Up

1	kcalorie	9.	lean	17.	essential	25.	social
2.	exchange	10.	hormonal	18.	22	26.	underweight
3.	heat	11.	3,500	19.	32	27.	cancer
4.	body	12.	fluid	20.	amount	28.	Infertility
5.	oxygen	13.	lean	21.	Intra-abdominal	29.	sleep apnea
6.	energy	14.	75	22.	central	30.	surgery
7.	basal	15.	25	23.	cancer	31.	chronic
8.	muscular	16.	weights	24.	waist		

Chapter Study Questions

1. Overfatness and underweight.
2. Hunger is the physiological need for food; appetite is the psychological desire for food. Satiation is the feeling of satisfaction and fullness that occurs during a meal and halts eating; satiety is a feeling of fullness after a meal. Hunger and appetite are likely to increase food intake. Satiation determines how much food is consumed during a meal. Satiety inhibits eating until the next meal; it determines how much time passes between meals. A high level of satiety is likely to decrease food intake between meals.
3. Basal metabolism: energy needed to maintain life when a body is at complete rest. Physical activity: voluntary movement of the skeletal muscles and support systems. Thermic effect of food: energy required to process food. See "How to Estimate Energy Requirements" box for specific equations.
4. Weight depends on frame size, lean mass, fat mass, and bone weight; body composition is more important than weight but is difficult to measure. Location of body fat is more important than quantity of body fat, fat around the waist and abdomen poses a greater risk than fat in other areas; and assessment techniques include: total body water, radioactive potassium count, near-infrared spectrophotometry, ultrasound, computer tomography, magnetic resonance imaging.
5. Defining a healthy body weight is a complex issue. Our society sets unrealistic ideals for body weight, especially for women. These ideals usually impair the health of those who attempt to adhere to them. Body composition is another important issue. Two people may weigh the same but have different amounts of muscle and fat. The presence of more lean mass is beneficial. Another factor is the distribution of body fat. Fat that accumulates within the abdomen poses a health risk.
6. Central obesity is excess fat on the abdomen and around the trunk of the body. It presents a greater risk to health than fat elsewhere on the body.
7. Increased risk of heart attacks, strokes, diabetes, high blood cholesterol, hypertension, surgery complications, gynecological problems, toxemia of pregnancy, certain types of cancer, arthritis, abdominal hernias, respiratory problems, gout, and accidents.

Sample Test Questions

1. d (p. 249)	9. a (p. 250)	17. b (p. 253)	25. d (p. 263)
2. c (p. 250)	10. b (p. 250)	18. d (p. 249-250)	26. a (p. 263)
3. a (p. 251)	11. b (p. 254)	19. c (p. 259)	27. d (p. 264)
4. c (p. 251-252)	12. b (p. 254)	20. b (p. 256)	28. b (p. 265)
5. d (p. 254)	13. d (p. 254)	21. a (p. 263-264)	29. c (p. 265)
6. c (p. 253)	14. d (p. 254-255)	22. d (p. 258-259)	30. c (p. 265)
7. c (p. 253)	15. a (p. 254)	23. a (p. 259)	
8. c (p. 256)	16. c (p. 252)	24. c (p. 259-260)	

Short Answer Questions

1. basal metabolism; physical activities

Problem Solving

1. 175 lb. divided by 2.2 lb./kg = 79 kg
   79 kg x 24 kcal/kg/day = 1896 kcal/day
2. 3500 kcalories in a pound of fat divided by 7 days per week = 500
   2500 – 500 = 2000 kcalories per day
3. 150 x 703 = 105450
   75^2 = 75 x 75 = 5625
   105450 divided by 5625 = 18.7
4. 132 x 703 = 92798
   66 x 66 = 4356
   92798 divided by 4356 = 21.3
5. 3000 x .65 = 1950 BMR kcalories per day
   3000 x .10 = 300 kcalories from thermic effect of food
   3000 x .25 = 750 physical activity kcalories
6. 24 divided by 0.152 = 158 pounds
   Alternative method: 24 = (wt. x 703) divided by (68^2)
   24 = (wt. x 703) divided by 4,624
   24 x 4,624 = wt x 703
   110,976 divided by 703 = wt. = 158 lb.
7. Refer to Table 8-2 for kcal/lb./minute expended on various activities.
   150 x 0.062 kcal/lb./min = 9.3 kcal/min
   9.3 kcal/min x 30 minutes = 279 kcalories
8. 170 x 0.103 kcal/lb./min = 17.51 kcal/min
   17.51 x 20 min = 350 kcalories
9. EER = 354 – (6.91 x 47) + 1.12 x [(9.36 x 59) + (726 x 1.68)]
   EER = 354 – 325 + 1.12 x [552 + 1219]
   EER = 354 – 325 + 1.12 x 1771
   EER = 354 – 325 + 1983 = 2012 kcalories/day
10. EER = [662 – (9.53 x 21)] + 1.25 x [(15.91 x 65) + (539.6 x 1.8)]
    EER = [662 – 200] + 1.25 x [1034 + 971]
    EER = 462 + 1.25 x 2005 = 462 + 2506 = 2968 kcalories/day

Crossword Puzzle

1. hydrodensitometry
2. thermogenesis
3. fatfold measure
4. hypothalamus

5. lean body mass
6. direct
7. indirect
8. BMI

9. body composition
10. basal metabolism

⑥ Chapter 9 – Weight Management: ⑥ Overweight, Obesity, and Underweight

Chapter Outline

I. Overweight
    A. Fat Cell Development
    B. Fat Cell Metabolism
    C. Set-Point Theory
II. Causes of Obesity
    A. Genetics
        1. Leptin
        2. Ghrelin
        3. Uncoupling Proteins
    B. Environment
        1. Overeating
        2. Physical Inactivity
III. Problems of Obesity
    A. Health Risks
        1. Overweight in Good Health
        2. Obese or Overweight with Risk Factors
        3. Obese or Overweight with Life Threatening Conditions
    B. Perceptions and Prejudices
        1. Social Consequences
        2. Psychological Problems
    C. Dangerous Interventions
        1. Fad Diets
        2. Weight-Loss Products
        3. Other Gimmicks
IV. Aggressive Treatments of Obesity
    A. Drugs
        1. Sibutramine
        2. Orlistat
        3. Other Drugs
    B. Surgery
V. Weight-Loss Strategies
    A. Eating Plans
        1. Be Realistic about Energy Intake
        2. Emphasize Nutritional Adequacy
        3. Eat Small Portions
        4. Lower Energy Density
        5. Remember Water
        6. Focus on Fiber
        7. Choose Fats Sensibly
        8. Watch for Other Empty kCalories

    B. Physical Activity
        1. Activity and Energy Expenditure
        2. Activity and Discretionary kCalorie Allowance
        3. Activity and Metabolism
        4. Activity and Body Composition
        5. Activity and Appetite Control
        6. Activity and Psychological Benefits
        7. Choosing Activities
        8. Spot Reducing
    C. Behavior and Attitude
        1. Become Aware of Behaviors
        2. Change Behaviors
        3. Personal Attitude
        4. Support Groups
    D. Weight Maintenance
    E. Prevention
    F. Public Health Programs
VI. Underweight
    A. Problems of Underweight
    B. Weight-Gain Strategies
        1. Energy-Dense Foods
        2. Regular Meals Daily
        3. Large Portions
        4. Extra Snacks
        5. Juice and Milk
        6. Exercising to Build Muscles
VII. The Latest and Greatest Weight-Loss Diet—Again
    A. The Diet's Appeal
    B. The Diet's Achievements
        1. Don't Count kCalories
        2. Satisfy Hunger
        3. Follow a Plan
        4. Limit Choices
    C. The Diet's Shortcomings
        1. Too Much Fat
        2. Too Much Protein
        3. Too Little Everything Else
    D. The Body's Perspective

Summing Up

The prevalence of overweight and obesity in the United States continues to (1)_____ dramatically. Overweight is especially high among women, the (2)_____, blacks and

(3)_____. The prevalence of overweight among children has also risen at an
(4)_____ rate. Obesity is so widespread that many refer to it as an (5)_____.

Excess body fat accumulates when people consistently take in more food (6)_____ than
they (7)_____. Obesity likely has many (8)_____ causes. Probable causes of
obesity include (9)_____ development, genetics, fat cell (10)_____,
(11)_____ theory, overeating, and (12)_____. Obesity is usually accompanied
by several (13)_____ problems.

Since causes of obesity vary widely, treatment must be (14)_____ and multifaceted.
Ineffective treatments include the use of (15)_____, amphetamines, and hormones.
(16)_____ is a justified approach in some specific cases of clinically severe obesity. The most
important component of successful treatment is the adoption of a balanced and nourishing low-kcalorie diet; fat loss
is greatly enhanced by regular (17)_____; and (18)_____ modification helps
these adaptive strategies to be maintained. Major criteria for success are a permanent change in
(19)_____ habits and (20)_____ of the goal weight over the long term.

Diet and behavior modification can also help the underweight person to gain weight. But the special case of
anorexia (21)_____ requires skilled professional attention. In all weight-control problems—
obesity, underweight, and anorexia—real success is achieved only when new, adaptive eating and coping
(22)_____ have permanently replaced the old ones.

Chapter Study Questions

1. Describe how body fat develops and suggest some reasons why it is difficult for an obese person to maintain weight loss.

2. What factors contribute to obesity?

3. List several aggressive ways to treat obesity and explain why such methods are not recommended for every overweight person.

4. Discuss reasonable dietary strategies suitable for achieving and maintaining a healthy body weight.

5. What are the benefits of increased physical activity in a weight-loss program?

6. Describe the behavioral strategies recommended for changing an individual's dietary habits. What role does personal attitude play?

7. Describe strategies for successful weight gain.

Chapter Glossary

- **bariatrics:** the field of medicine specializing in the treatment of obesity.
- **behavior modification:** the changing of behavior by the manipulation of antecedents (cues or environmental factors that trigger behavior), the behavior itself, and consequences (the penalties or rewards attached to behavior).
- **brown adipose tissue:** masses of specialized fat cells packed with pigmented mitochondria that produce heat instead of ATP.
- **cellulite:** supposedly, a lumpy form of fat; actually, a fraud. Fatty areas of the body may appear lumpy when the strands of connective tissue that attach the skin to underlying muscles pull tight where the fat is thick. The fat itself is the same as fat anywhere else in the body. If the fat in these areas is lost, the lumpy appearance disappears.
- **clinically severe obesity:** a BMI of 40 or greater or a BMI of 35 or greater with additional medical problems. A less preferred term used to describe the same condition is morbid obesity.
- **epidemic:** the appearance of a disease (usually infectious) or condition that attacks many people at the same time in the same region.
- **fad diets:** popular eating plans that promise quick weight loss. Most fad diets severely limit certain foods or overemphasize others (for example, never eat potatoes or pasta or eat cabbage soup daily).
- **gene pool:** all the genetic information of a population at a given time.
- **ghrelin:** a protein produced by the stomach cells that enhances appetite and decreases energy expenditure.
- **hyperplastic obesity:** obesity due to an increase in the *number* of fat cells.
- **hypertrophic obesity:** obesity due to an increase in the *size* of fat cells.
- **leptin**: a protein produced by fat cells under direction of the *ob* gene that decreases appetite and increases energy expenditure; sometimes called the *ob protein*.
- **lipotoxicity**: the adverse effects of fat in nonadipose tissues.
- **orlistat:** a drug used in the treatment of obesity that inhibits the absorption of fat in the GI tract, thus limiting kcaloric intake.
- **serotonin:** a neurotransmitter important in sleep regulation, appetite control, and sensory perception, among other roles. Serotonin is synthesized in the body from the amino acid tryptophan with the help of vitamin B_6.
- **set point:** the point at which controls are set (for example, on a thermostat). The set-point theory that relates to body weight proposes that the body tends to maintain a certain weight by means of its own internal controls.
- **sibutramine:** a drug used in the treatment of obesity that slows the reabsorption of serotonin in the brain, thus suppressing appetite and creating a feeling of fullness.

- **successful weight-loss maintenance:** achieving a weight loss of at least 10 percent of initial body weight and maintaining the loss for at least one year.
- **weight management:** maintaining body weight in a healthy range by preventing gradual weight gain over time and losing weight if overweight.

Sample Test Questions

Select the best answer for each question.

1. The prevalence of overweight and obesity continues to:
   a. decrease.
   b. increase.
   c. stay constant.
   d. rise and fall.

2. A person with a BMI of 26 is:
   a. underweight.
   b. at a healthy weight.
   c. overweight.
   d. obese.

3. Obesity is viewed as:
   a. a minor health risk.
   b. a condition that improves physical appearance.
   c. a pandemic.
   d. an epidemic.

4. Obesity due to an increase in the size of fat cells is:
   a. hypertrophic obesity.
   b. hyperplastic obesity.
   c. lipotoxic obesity.
   d. lipase obesity.

5. Obesity due to an increase in the number of fat cells is called:
   a. hypertrophic obesity.
   b. hyperplastic obesity.
   c. epidemic obesity.
   d. none of the above

6. Which enzyme promotes efficient fat storage in both fat and muscle cells?
   a. protease
   b. brown fat lipase
   c. leptin protease
   d. lipoprotein lipase

7. The adverse effects of fat in nonadipose areas are called:
   a. lipase toxicity.
   b. nonadipose toxicity.
   c. lipotoxicity.
   d. organ degeneration.

8. _____ suggests that the body tends to maintain a certain weight by internal controls.
   a. LPL activity theory
   b. Lepos theory
   c. Leptin theory
   d. Set-point theory

9. The obesity gene is called the:
   a. *ob* gene.
   b. set point.
   c. *leptin* gene.
   d. *Gap* gene.

10. When leptin levels are high:
    a. leptin acts on the hypothalamus.
    b. appetite is reduced.
    c. energy expenditure is slowed.
    d. a and b
    e. a, b, and c

11. Which protein triggers the desire to eat?
    a. ghrelin
    b. leptin
    c. protease
    d. lipase

12. Adverse reactions to fad diets can include:
    a. headaches and dizziness.
    b. nausea.
    c. death.
    d. a and b
    e. a, b, and c

13. A drug that slows the reabsorption of serotonin in the brain is called:
    a. orlistat.
    b. benzocaine.
    c. phenylpropanolamine.
    d. sibutramine.

14. Which of the following is accurate regarding chromium?
    a. It affects body composition and metabolism.
    b. It permits the achievement of higher levels of energy expenditure and thus weight loss.
    c. It is ineffective for eliminating body fat.
    d. It is a safe method of weight loss.

15. A drug that inhibits fat absorption in the GI tract is:
    a. sibutramine.
    b. orlistat.
    c. ephedrine.
    d. yohimbine.

16. Methods to induce weight loss by interfering with the amount of food consumed include all of the following except
    a. liposuction.
    b. gastric banding.
    c. reducing the size of the stomach.
    d. gastric bypass.

17. Which of the following does not describe the controversies involved in obesity treatment?
    a. Treating obesity is a simple task.
    b. Everyone cannot achieve thinness.
    c. Overweight people face discrimination.
    d. Self-esteem is harmed from repeated weight loss and gain.

18. A safe rate for weight loss is:
    a. ½ to 2 lb per week.
    b. 3 to 5 lb per week.
    c. 10% of body weight over 2 months.
    d. 15% of body weight over 3 months.

19. What is the best approach to weight loss?
    a. Avoid foods containing carbohydrates.
    b. Reduce daily energy intake and increase energy expenditure.
    c. Eliminate all fats from the diet and decrease water intake.
    d. Greatly increase protein intake to prevent body protein loss.

20. To lose fat efficiently while retaining lean tissue, a person needs a deficit of _____ kcalories per day.
    a. 500
    b. 100
    c. 150
    d. 200

21. Fiber-rich foods usually offer:
    a. high energy density.
    b. high fat amounts.
    c. high nutrient density.
    d. low satiety.

22. A benefit of fat in a weight control program is its:
    a. low energy density.
    b. strong satiation effect.
    c. high nutrient density.
    d. contribution of essential nutrients.

23. A tip for weight loss is:
    a. cut out all fats in the diet.
    b. increase your daily energy intake.
    c. avoid all foods containing carbohydrates.
    d. increase your water intake.

24. Regular physical activity often results in all of the following except:
    a. lower energy expenditure.
    b. raised BMR.
    c. improved appetite control.
    d. stress reduction.

25. The *Dietary Guidelines* recommend _____ minutes of activity on most days of the week.
    a. 30
    b. 45
    c. 60
    d. 90 or more

26. Which of the following statements is true regarding exercise?
    a. Specific exercises can remove fat from certain targeted body parts.
    b. Exercise can assist in weight loss by burning fat.
    c. Exercise is only beneficial if it is done for 1 hour or more.
    d. Older people should not engage in physical activity.

27. To help maximize the long-term success of a person's weight-loss program, which of the following personal attitudes should be encouraged in the individual?
    a. Openness to examining emotional health status and whether stress triggers overeating
    b. Viewing the body realistically as being very fat rather than thin
    c. Refraining from expressing overconfidence in ability to lose weight
    d. Accepting that lack of exercising is a part of the lifestyle of most overweight people

28. All of the following are behavior modifications for losing weight except:
    a. shopping only when not hungry.
    b. eating only in one place and in one room.
    c. watching television only when not eating.
    d. taking smaller portions of food but eating everything quickly.

29. Which of the following would probably not be part of a successful program of weight gain in an underweight individual?
    a. engaging in physical exercise to build muscle tissue
    b. consuming energy-dense foods
    c. consuming energy-dense beverages
    d. consuming a large number of small meals

30. When working on weight gain, at first it takes an excess of approximately _____ kcalories per day above normal energy needs.
    a. 300
    b. 500
    c. 550-600
    d. 750-800

Problem Solving

1. Your friend has calculated 1800 kcalories/day as appropriate for a healthy rate of weight loss. Calculate the number of kcalories from carbohydrate, protein and fat that are recommended for a weight-loss diet.

2. You have decided to consume 2000 kcalories per day for weight control purposes. How many kcalories should you consume from saturated, monounsaturated, and polyunsaturated fatty acids?

3. Calculate the energy density of a food item which weighs 75 grams and delivers 35 kcalories.

4. Calculate the energy density of a food item which weighs 60 grams and delivers 100 kcalories.

5. A person's EER is 4000 kcalories per day, current weight is 210, and desired weight is 175. This person plans to lose 1 pound of fat a week. How many kcalories should be consumed and (if compliant) how long will it take to achieve the desired weight?

6. A mixed creamy alcoholic beverage contains 2 grams of alcohol, 60 grams of carbohydrate, 20 grams of protein and 40 grams of fat. Calculate the kcalories that this beverage can deliver.

7. A person is overweight (350 pounds) and desires to lose 10% of his body weight over 6 months. Calculate the amount of total desired weight loss and amount of desired weight loss per month.

8. In question #7, what is the total kcalorie intake restriction required to reach the weight loss goal? If the weight loss time period is 27 weeks, what are the weekly and daily kcalorie restrictions to achieve this goal?

9. A person has a BMI of 15.5 and desires to gain weight. Do you support the person's desire to gain weight based on the categorization of the BMI? If the person's EER is 2100, to start, how many kcalories should be consumed per day to support weight gain of 1 pound per week?

10. A person has a BMI of 27 and needs to lose weight. How many kcalories per day reduction from usual intake should this person consume? If the EER is 4000 kcalories, how many kcalories from carbohydrate, protein and fat should be consumed?

Crossword Puzzle

Complete this crossword puzzle by Mary A. Wyandt, Ph.D., CHES.

Across:		Down:	
2.	obesity in which the BMI is of 40 or greater or 100 pounds or more overweight for an average adult	1.	a protein produced by the stomach cells that enhances appetite and decreases energy expenditure
4.	all the genetic information of a population at a given time	3.	adverse effects of fat in non-adipose tissues
7.	masses of specialized fat cells packed with pigmented mitochondria that produce heat instead of ATP	5.	an environment; referring to easy access to and overabundance of high-fat, high-kcalorie foods in our society
9.	the point at which controls are set	6.	supposedly a lumpy form of fat; actually a fraud
10.	a protein produced by fat cells under direction of the ob gene that decreases appetite and increases energy expenditure	8.	a neurotransmitter important in sleep regulation, appetite control and sensory perception

Chapter 9 Answer Key

Summing Up

1. rise
2. poor
3. Hispanics
4. alarming
5. epidemic
6. energy
7. spend
8. interrelated
9. fat cell
10. metabolism
11. set-point
12. inactivity
13. health
14. individualized
15. diuretics
16. Surgery
17. exercise
18. behavior
19. eating
20. maintenance
21. nervosa
22. behaviors

Chapter Study Questions

1. Body fat develops when fat cells increase in number and size. Prevention of excess weight gain depends on maintaining a reasonable number of fat cells; when an obese person loses weight, the fat cells shrink, but their number does not decrease. According to set-point theory, the body attempts to return to the original weight, or its set point.
2. Genetics (leptin, ghrelin, uncoupling problems, fat cell metabolism, set point), overeating, and inactivity.
3. Drugs including sibutramine that suppresses appetite, orlistat that blocks dietary fat digestion, and off-label drugs that may cause weight loss incidental to treating other conditions. Surgery including gastric surgery to limit food intake by reducing the capacity of the stomach and liposuction to extract some fat deposits. Reasons these are not recommended for everyone include risks and results are not always effective. Risks of sibutramine include dry mouth, headache, constipation, rapid heart rate, and high blood pressure. Risks of orlistat include gas, frequent bowel movements, and reduced absorption of fat-soluble vitamins. Risks of gastric surgery include nausea, vomiting, dehydration and certain vitamin deficiencies. Liposuction can cause complications. Obese people should be carefully screened prior to utilizing an aggressive method and be counseled regarding the need to comply with dietary advice.
4. Eating plans based on realistic energy intake, nutritional adequacy, physical activity, making small behavior modification changes, and support groups.
5. Physical activity increases BMR, helps control appetite, and provides psychological benefits.
6. Record food intake using a food record to help become aware of behaviors, focus on learning desired eating and exercise behaviors and eliminating unwanted behaviors, do not grocery shop when hungry, eat slowly, and exercise when watching television. A personal attitude of sound emotional health supports positive behavior change.
7. Eat energy-dense foods to provide an excess of 750-800 kcal per day, eat regular meals daily, eat large portions, eat extra snacks between meals, drink plenty of juice and milk, and exercise to build muscle.

Sample Test Questions

1. b (p. 281)
2. c (p. 281)
3. d (p. 282)
4. a (p. 283)
5. b (p. 283)
6. d (p. 282)
7. c (p. 283)
8. d (p. 283)
9. a (p. 284)
10. d (p. 284)
11. a (p. 285)
12. e (p. 290)
13. d (p. 292)
14. c (p. 291)
15. b (p. 292)
16. a (p. 292-293)
17. a (p. 289)
18. a (p. 295)
19. b (p. 295)
20. a (p. 296)
21. c (p. 296-297)
22. d (p. 298)
23. d (p. 297)
24. a (p. 300-301)
25. c (p. 300)
26. b (p. 300-302)
27. a (p. 304)
28. d (p. 303-304)
29. d (p. 308)
30. d (p. 307)

Problem Solving

1. 1800 x .30 = 540 kcal from fat
   1800 x .55 = 990 kcal from carbohydrate
   1800 x .15 = 270 kcal from protein

2. $2000 \times .08 = 160$; $2000 \times .10 = 200$; 160-200 kcal from saturated fatty acids
   $2000 \times .15 = 300$; up to 300 kcal from monounsaturated fatty acids
   $2000 \times .10 = 200$; up to 200 from polyunsaturated fatty acids
3. 35 divided by 75 = 0.47 kcal/g
4. 100 divided by 60 = 1.67 kcal/g
5. $4000 - 500$ kcalories = 3500 kcalories/day
   $210 - 175 = 35$ weeks
6. $2 \times 7 = 14$ kcalories from alcohol
   $60 \times 4 = 240$ kcalories from carbohydrate
   $20 \times 4 = 80$ kcalories from protein
   $40 \times 9 = 360$ kcalories from fat
   $14 + 240 + 80 + 360 = 694$ total kcalories
7. $350 \times .10 =$ desires to lose 35 pounds over 6 months. 35 divided by 6 months = 5.833 lb. per month.
8. 35 pounds $\times$ 3500 = 122,500 total kcalorie restriction
   122,500 divided by 27 weeks = 4537 fewer kcal per week
   4537 divided by 7 = 648 fewer kcal per day
9. The BMI of 15.5 is less than 18.5 so the person is classified as underweight. Weight gain in this case is appropriate for good health. To start, $2100 + 800 = 2900$ kcalories per day to support weight gain of 1 pound per week.
10. A person with a BMI of 27 should consume 300 to 500 kcalories per day less than his or her usual intake.
    $4000 - 500 = 3500$ kcalories per day for weight loss
    $3500 \times .55 = 1925$ kcalories from carbohydrate
    $3500 \times .15 = 525$ kcalories from protein
    $3500 \times .30 = 1050$ kcalories from fat

Crossword Puzzle

1. ghrelin
2. clinically severe
3. lipotoxicity
4. gene pool
5. toxic food
6. cellulite
7. brown adipose tissue
8. serotonin
9. set point
10. leptin

⑤ Chapter 10 – The Water- Soluble Vitamins: ⑤ B Vitamins and Vitamin C

Chapter Outline

I. The Vitamins—An Overview
   A. Bioavailability
   B. Precursors
   C. Organic Nature
   D. Solubility
   E. Toxicity
II. The B Vitamins—As Individuals
   A. Thiamin
      1. Thiamin Recommendations
      2. Thiamin Deficiency and Toxicity
      3. Thiamin Food Sources
   B. Riboflavin
      1. Riboflavin Recommendations
      2. Riboflavin Deficiency and Toxicity
      3. Riboflavin Food Sources
   C. Niacin
      1. Niacin Recommendations
      2. Niacin Deficiency
      3. Niacin Toxicity
      4. Niacin Food Sources
   D. Biotin
      1. Biotin Recommendations
      2. Biotin Deficiency and Toxicity
      3. Biotin Food Sources
   E. Pantothenic Acid
      1. Pantothenic Acid Recommendations
      2. Pantothenic Acid Deficiency and Toxicity
      3. Pantothenic Acid Food Sources
   F. Vitamin B_6
      1. Vitamin B_6 Recommendations
      2. Vitamin B_6 Deficiency
      3. Vitamin B_6 Toxicity
      4. Vitamin B_6 Food Sources
   G. Folate
      1. Folate Recommendations
      2. Folate and Neural Tube Defects
      3. Folate and Heart Disease
      4. Folate and Cancer
      5. Folate Deficiency
      6. Folate Toxicity
      7. Folate Food Sources
   H. Vitamin B_{12}
      1. Vitamin B_{12} Recommendations
      2. Vitamin B_{12} Deficiency and Toxicity
      3. Vitamin B_{12} Food Sources

I. Non-B Vitamins
   1. Choline
   2. Inositol and Carnitine
   3. Vitamin Impostors
III. The B Vitamins—In Concert
   A. B Vitamin Roles
   B. B Vitamin Deficiencies
   C. B Vitamin Toxicities
   D. B Vitamin Food Sources
IV. Vitamin C
   A. Vitamin C Roles
      1. As an Antioxidant
      2. As a Cofactor in Collagen Formation
      3. As a Cofactor in Other Reactions
      4. In Stress
      5. As a Cure for the Common Cold
      6. In Disease Prevention
   B. Vitamin C Recommendations
   C. Vitamin C Deficiency
   D. Vitamin C Toxicity
   E. Vitamin C Food Sources
V. Vitamin and Mineral Supplements
   A. Arguments for Supplements
      1. Correct Overt Deficiencies
      2. Support Increased Nutrient Needs
      3. Improve Nutrition Status
      4. Improve the Body's Defenses
      5. Reduce Disease Risks
      6. Who Needs Supplements?
   B. Arguments against Supplements
      1. Toxicity
      2. Life-Threatening Misinformation
      3. Unknown Needs
      4. False Sense of Security
      5. Other Invalid Reasons
      6. Bioavailability and Antagonistic Actions
   C. Selection of Supplements
      1. Form
      2. Contents
      3. Misleading Claims
      4. Cost
   D. Regulation of Supplements

Summing Up

The B vitamins serve as (1)_____ assisting many enzymes in the body. Thiamin, (2)_____, niacin, and pantothenic acid are especially important in the glucose-to-energy pathway; they are active in the coenzymes (3)_____, FAD, NAD+, and (4)_____, respectively.

Vitamin B$_6$ facilitates (5)_____ acid transformations and thus protein metabolism; (6)_____ is involved in pathways leading to the synthesis of new cells, and (7)_____ in the release of folate in its active form. (8)_____ is involved in lipid synthesis.

A lack of thiamin causes (9)_____; a lack of niacin (unless compensated for by its amino acid precursor tryptophan) causes (10)_____; a lack of the intrinsic factor for vitamin B$_{12}$ causes (11)_____ anemia. Human deficiencies have also been observed for riboflavin, vitamin B$_6$, and folate.

(12)_____ is widely distributed in foods, but no food contributes a great amount of it. (13)_____ is concentrated in milk and meats. (14)_____ is found wherever protein is found and can also be made from the amino acid tryptophan. Vitamin B$_6$ is most abundant in (15)_____, vitamin B$_{12}$ is found only in (16)_____ products, and (17)_____ is supplied best by green, leafy vegetables. Any diet plan that includes moderate amounts of all these foods ensures probable (18)_____ for these nutrients.

Vitamin C acts as an (19)_____, helping to maintain iron in its reduced (20)_____ form and thus cooperating with enzymes that require this form of iron as a (21)_____.

Vitamin C also helps regulate the overall oxidation-reduction equilibrium of the body (22)_____ and fluids. Vitamin C promotes the formation of the protein (23)_____. It is involved in the metabolism of several (24)_____. Deficiency of vitamin C causes (25)_____. Scurvy is prevented by the daily intake of only (26)_____ milligrams of vitamin C. Recommended daily intakes of vitamin C range from (27)_____ milligrams. Toxic effects of megadoses (3 to 10 grams) have been reported. The best food sources of vitamin C are the (28)_____ fruits, strawberries and cantaloupe, broccoli and other members of the cabbage family, and greens.

Chapter Study Questions

1. How do the vitamins differ from the energy nutrients?

2. Describe some general differences between fat-soluble and water-soluble vitamins.

3. Which B vitamins are involved in energy metabolism? Protein metabolism? Cell division?

4. For thiamin, riboflavin, niacin, biotin, pantothenic acid, vitamin B_6, folate, vitamin B_{12}, and vitamin C, state:
   • Its chief function in the body.
   • Its characteristic deficiency symptoms.
   • Its significant food sources.

5. What is the relationship of tryptophan to niacin?

6. Describe the relationship between folate and vitamin B_{12}.

7. What risks are associated with high doses of niacin? Vitamin B_6? Vitamin C?

Chapter Glossary

• **anemia:** literally, "too little blood." Anemia is any condition in which too few red blood cells are present, or the red blood cells are immature (and therefore large) or too small or contain too little hemoglobin to carry the normal amount of oxygen to the tissues. It is not a disease itself but can be a symptom of many different disease conditions, including many nutrient deficiencies, bleeding, excessive red blood cell destruction, and defective red blood cell formation.

- **antagonist:** a competing factor that counteracts the action of another factor. When a drug displaces a vitamin from its site of action, the drug renders the vitamin ineffective and thus acts as a vitamin antagonist.
- **antioxidant:** a substance in foods that significantly decreases the adverse effects of free radicals on normal physiological functions in the human body.
- **antiscorbutic factor:** the original name for vitamin C.
- **ariboflavinosis**: riboflavin deficiency causing inflammation of the membranes of the mouth, skin, eyes, and GI tract.
- **ascorbic acid:** one of the two active forms of vitamin C. Many people refer to vitamin C by this name.
- **atrophic gastritis:** chronic inflammation of the stomach accompanied by a diminished size and functioning of the mucous membrane and glands.
- **avidin:** the protein in egg whites that binds biotin.
- **beriberi:** the thiamin-deficiency disease.
- **bioavailability:** the rate at and the extent to which a nutrient is absorbed and used.
- **biotin:** a B vitamin that functions as a coenzyme in metabolism.
- **carnitine:** a nonessential, nonprotein amino acid made in the body from lysine that helps transport fatty acids across the mitochondrial membrane. Carnitine supposedly "burns" fat and spares glycogen during endurance events, but in reality it does neither.
- **carpal tunnel syndrome:** a pinched nerve at the wrist, causing pain or numbness in the hand. It is often caused by repetitive motion of the wrist.
- **cheilosis**: a condition of reddened lips with cracks at the corners of the mouth that is a symptom commonly seen in B vitamin deficiencies.
- **choline:** a nitrogen-containing compound found in foods and made in the body from the amino acid methionine. Choline is part of the phospholipid lecithin and the neurotransmitter acetylcholine.
- **cofactor:** a small inorganic or organic substance that facilitates the action of an enzyme.
- **dietary folate equivalents (DFE):** the amount of folate available to the body from naturally occurring sources, fortified foods, and supplements, accounting for differences in the bioavailability from each source.
- **false negative:** a test result indicating that a condition is not present (negative) when in fact it is present (therefore false).
- **false positive:** a test result indicating that a condition is present (positive) when in fact it is not (therefore false).
- **folate:** a B vitamin; also known as *folic acid, folacin,* or *pteroylglutamic acid (PGA).* The coenzyme forms are DHF (dihydrofolate) and THF (tetrahydrofolate).
- **free radicals:** unstable molecules with one or more unpaired electrons.
- **glossitis:** an inflammation of the tongue that is a symptom commonly seen in B vitamin deficiencies.
- **gluconeogenesis**: the synthesis of glucose from noncarbohydrate sources such as amino acids or glycerol.
- **histamine:** a substance produced by cells of the immune system as part of a local immune reaction to an antigen; participates in causing inflammation.
- **inositol:** a nonessential nutrient that can be made in the body from glucose. Inositol is a part of cell membrane structures.
- **intrinsic factor:** a glycoprotein (a protein with short polysaccharide chains attached) manufactured in the stomach that aids in the absorption of vitamin B_{12}.
- **macrocytic anemia:** large-cell anemia; also known as *megaloblastic anemia.*
- **megaloblastic anemia:** large-cell anemia; also known as *macrocytic anemia.*
- **neural tube defects:** malformations of the brain, spinal cord, or both during embryonic development that often result in lifelong disability or death.
- **niacin:** a B vitamin. The coenzyme forms are NAD (nicotinamide adenine dinucleotide) and NADP (the phosphate form of NAD). Niacin can be eaten preformed or made in the body from its precursor, tryptophan, one of the amino acids.
- **niacin equivalents (NE):** the amount of niacin present in food, including the niacin that can theoretically be made from its precursor, tryptophan, present in the food.
- **niacin flush:** a temporary burning, tingling, and itching sensation that occurs when a person takes a large dose of nicotinic acid; often accompanied by a headache and reddened face, arms, and chest.

- **oxidative stress:** a condition in which the production of oxidants and free radicals exceeds the body's ability to handle them and prevent damage.
- **pantothenic acid:** a B vitamin. The principal active form is part of coenzyme A, called "CoA" throughout Chapter 7.
- **pellagra:** the niacin-deficiency disease.
- **pernicious anemia:** a blood disorder that reflects a vitamin B_{12} deficiency caused by lack of intrinsic factor and characterized by abnormally large and immature red blood cells. Other symptoms include muscle weakness and irreversible neurological damage.
- **pharmacological effect:** describes the effect of a nutrient when a large dose (levels commonly available only from supplements) overwhelms some body system and acts like a drug.
- **physiological effect:** describes the effect of a nutrient when a normal dose of a nutrient (levels commonly found in foods) provides a normal blood concentration.
- **precursors:** substances that precede others; with regard to vitamins, compounds that can be converted into active vitamins; also known as *provitamins*.
- **riboflavin:** a B vitamin. The coenzyme forms are FMN (flavin mononucleotide) and FAD (flavin adenine dinucleotide).
- **scurvy:** the vitamin C–deficiency disease.
- **serotonin:** a neurotransmitter important in appetite control, sleep regulation, and sensory perception, among other roles; it is synthesized in the body from the amino acid tryptophan with the help of vitamin B_6.
- **thiamin:** a B vitamin. The coenzyme form is TPP (thiamin pyrophosphate).
- **vitamin B_6:** a family of compounds—pyridoxal, pyridoxine, and pyridoxamine. The primary active coenzyme form is PLP (pyridoxal phosphate).
- **vitamin B_{12}:** a B vitamin characterized by the presence of cobalt. The active forms of coenzyme B_{12} are methylcobalamin and deoxyadenosylcobalamin.
- **Wernicke-Korsakoff Syndrome:** severe thiamin deficiency in alcohol abusers; symptoms include disorientation, loss of short-term memory, jerky eye movements, and staggering gait.

Sample Test Questions

Select the best answer for each question.

1. Vitamins are:
   a. organic.
   b. inorganic.
   c. essential nutrients required in small amounts.
   d. a and c
   e. b and c

2. Vitamins differ from carbohydrates, protein, and fat in the following way(s):
   a. function.
   b. structure.
   c. amounts required.
   d. a and b only
   e. a, b, and c

3. Vitamins are similar to carbohydrates, protein and fat in the following way(s):
   a. function.
   b. vital to life.
   c. organic.
   d. a and b
   e. b and c

4. A measure of the amount of vitamins absorbed and used by the body is called:
   a. bioavailability.
   b. dissolvability.
   c. organic elements.
   d. utilization.

5. The body's ability to use vitamins is determined by:
   a. time of transit through the GI tract.
   b. previous nutrient intake.
   c. method of food preparation.
   d. a and b
   e. a, b and c

6. Some vitamins that are present in inactive forms are called:
   a. synthetic.
   b. precursors.
   c. organic.
   d. elemental.

7. To minimize nutrient losses in fruits and vegetables:
   a. keep them on kitchen counters.
   b. use long cooking times for vegetables.
   c. store fruits and vegetables in airtight containers.
   d. cook vegetables in large amounts of water.

8. Which nutrients are hydrophilic?
   a. all vitamins
   b. B vitamins
   c. fat-soluble vitamins
   d. lipids

9. Water-soluble vitamins consumed in excess of need may be:
   a. malabsorbed.
   b. beneficial.
   c. harmful.
   d. a and b
   e. a and c

10. B vitamins:
   a. are water soluble.
   b. function as coenzymes.
   c. are stored in the adipose tissue.
   d. are absorbed into the lymph.
   e. a and b

11. A dietary deficiency of B vitamins can cause:
   a. negative nitrogen balance.
   b. impairment of energy metabolism.
   c. impaired iron utilization.
   d. reduced thyroid production.

12. A deficiency of thiamin produces the disease:
   a. rickets.
   b. pellagra.
   c. beriberi.
   d. scurvy.

13. Which of the following deficiencies is observed in alcohol abusers?
   a. thiamin
   b. niacin
   c. folate
   d. vitamin B_6

14. Which of the following vitamins is most readily destroyed by ultraviolet light?
   a. niacin
   b. thiamin
   c. riboflavin
   d. ascorbic acid

15. The dietary need for _____ is influenced by the presence of the amino acid tryptophan in the diet.
   a. thiamin
   b. pantothenic acid
   c. niacin
   d. biotin
   e. riboflavin

16. Two coenzyme forms of niacin are:
   a. FMN and FAD.
   b. NAD and FAD.
   c. NADP and TPP.
   d. FAD and TPP.
   e. NADP and NAD.

17. The disease called pellagra is related to a:
    a. dietary deficiency of riboflavin.
    b. low-protein diet high in corn products.
    c. diet high in polished rice.
    d. a and b
    e. a and c

18. Which of the following is highest in riboflavin content per serving?
    a. butter
    b. milk
    c. bread
    d. apple
    e. egg

19. If your skin feels flushed after taking a vitamin pill you may have overdosed on:
    a. niacin.
    b. thiamin.
    c. riboflavin.
    d. folate.

20. Pantothenic acid is a part of the structure of:
    a. pyruvate.
    b. glucose.
    c. cobalamin.
    d. glycogen.
    e. coenzyme A.

21. Which of the following statements is **true** about vitamin B_6?
    a. It is stored exclusively in fat tissue.
    b. It enhances athletic performance, especially endurance.
    c. It cures premenstrual syndrome.
    d. Large doses may cause irreversible nerve damage.

22. Often a function of one vitamin depends on the presence of another. This interdependence is shown in which two vitamins?
    a. vitamin B_{12} and folate
    b. vitamin A and vitamin C
    c. vitamin B_{12} and niacin
    d. folate and vitamin C

23. Vitamin B_{12} is different from other B vitamins because:
    a. it may be synthesized from a certain amino acid.
    b. it requires a carrier to be transported from the intestinal tract to the bloodstream.
    c. deficiencies of vitamin B_{12} never occur in human beings.
    d. it is found only in plant foods.

24. _____ may be lacking in the diet of strict vegetarians.
    a. Vitamin B_{12}
    b. Thiamin
    c. Riboflavin
    d. Niacin
    e. Biotin

25. Folate has proven critical in reducing the risks of:
    a. learning disabilities.
    b. alcoholism.
    c. neural tube defects.
    d. ADD.

26. The best food source of folate, among the following, is:
    a. spinach.
    b. milk.
    c. coffee.
    d. ice cream.

27. The vitamin C deficiency disease is:
    a. ascorbic acidosis.
    b. scurvy.
    c. pellagra.
    d. beriberi.

28. Most of the symptoms of a vitamin C deficiency are caused by:
    a. anemia.
    b. failure to maintain integrity of blood vessels.
    c. inactivity of intestinal bacteria.
    d. decreased utilization of protein.

29. The amount of vitamin C needed to prevent scurvy is about:
    a. 10 milligrams a day.   c. 60 milligrams a day.
    b. 35 milligrams a day.   d. 100 milligrams a day.

30. Which of the following is lowest in vitamin C?
    a. strawberries   d. cauliflower
    b. potatoes   e. orange juice
    c. milk

Short Answer Questions

1. Characteristics of water-soluble vitamins are:

    a.   c.

    b.   d.

2. The names of the water-soluble vitamins are:

    a.   f.

    b.   g.

    c.   h.

    d.   i.

    e.

3. Distinguish between these types of deficiencies:

    a. primary:

    b. secondary:

Problem Solving

1. A person whose RDA for protein is 45 grams consumes 75 grams of protein in a day. How many niacin equivalents does this represent?

2. What percentage of the RDA for vitamin C does a female nonsmoker receive from ½ cup of broccoli?

Crossword Puzzle

Complete this crossword puzzle by Mary A. Wyandt, Ph.D., CHES.

Across:	Down:
2. a substance in foods that significantly decreases the adverse effects of free radicals on normal physiological functions in the human body	1. a B vitamin; the coenzyme forms are FMN and FAD
6. a nitrogen-containing compound found in foods and made in the body from an amino acid	3. a B vitamin; the coenzyme form is TPP
8. a B vitamin that functions as a coenzyme in the metabolism of carbohydrates and fats	4. a nonessential nutrient that can be made in the body from glucose; it is used in cell membranes
9. a B vitamin, which can be eaten preformed or can be made in the body from its precursor, tryptophan	5. a nonessential nutrient made in the body from the amino acid lysine
10. a B vitamin; also known as folic acid	7. the protein in egg whites that binds biotin

❧ Chapter 10 Answer Key ❧

Summing Up

1. coenzymes
2. riboflavin
3. TPP
4. coA
5. amino
6. folate
7. vitamin B_{12}

8. Biotin
9. beriberi
10. pellagra
11. pernicious
12. Thiamin
13. Riboflavin
14. Niacin

15. meats
16. animal
17. folate
18. adequacy
19. antioxidant
20. (ferrous iron)
21. cofactor

22. cells
23. collagen
24. amino acids
25. scurvy
26. 10
27. 30 to 75
28. citrus

Chapter Study Questions

1. They differ in structure, function, and food contents.
2. Water-soluble vitamins are: carried in the blood, excreted in the urine, needed in frequent, small doses, and unlikely to reach toxic levels in the body. Fat-soluble vitamins are: absorbed into the lymph and carried in the blood by protein carriers, stored in body fat, needed in periodic doses, and more likely to be toxic when consumed in excess of needs.
3. B vitamins involved in energy metabolism: thiamin, riboflavin, niacin, biotin, and pantothenic acid. B vitamin involved in protein metabolism: vitamin B_6. B vitamins involved in cell division: folate and vitamin B_{12}.
4. See text for respective summary tables for each nutrient.
5. Tryptophan can be converted to niacin in the body: 60 mg tryptophan = 1 mg niacin.
6. Vitamin B_{12} and folate's roles intertwine because each depends on the other for activation. Folate is part of a coenzyme that helps convert vitamin B_{12} to one of its coenzyme forms and helps synthesize DNA; vitamin B_{12} removes a methyl group to activate the folate coenzyme.
7. Niacin: diarrhea, heartburn, nausea, ulcer irritation, vomiting, fainting, dizziness, painful flush and rash, excessive sweating, liver damage, and low blood pressure. Vitamin B_6: depression, fatigue, headaches, nerve damage, muscle weakness, numbness, and bone pain. Vitamin C: nausea, abdominal cramps, diarrhea, false results in diabetes tests, interference in drug effectiveness, hemolytic anemia, stone formation, and rebound scurvy.

Sample Test Questions

1. d (p. 323)
2. e (p. 323)
3. e (p. 324)
4. a (p. 324)
5. e (p. 324)
6. b (p. 324)
7. c (p. 324)
8. b (p. 324)

9. c (p. 325)
10. e (p. 326)
11. b (p. 326)
12. c (p. 327)
13. a (p. 327)
14. c (p. 330)
15. c (p. 331)
16. e (p. 331)

17. b (p. 332)
18. b (p. 329, 331)
19. a (p. 332)
20. e (p. 335)
21. d (p. 336)
22. a (p. 338, 342)
23. b (p. 343)
24. a (p. 344)

25. c (p. 338)
26. a (p. 341)
27. b (p. 350)
28. b (p. 353)
29. a (p. 353)
30. c (p. 334)

Short Answer Questions

1. carried in bloodstream; excreted in urine; needed in frequent small doses; less likely to be toxic
2. thiamin, riboflavin, vitamin B_6, niacin, folate, vitamin B_{12}, biotin, pantothenic acid, vitamin C
3. primary = inadequate intake; secondary = impaired absorption or metabolism, or excessive excretion

Problem Solving

1. 75 g - 45 g = 30 g protein
    30 g divided by 100 = 0.3 g tryptophan
    0.3 g x 1000 = 300 mg tryptophan
    300 mg divided by 60 = 5 mg niacin equivalents
2. ½ cup broccoli = 50 mg vitamin C; RDA for nonsmoking female adults = 75 mg
    50 mg divided by 75 mg = 67%

Crossword Puzzle

1.	riboflavin	3.	thiamin	5.	carnitine	7.	avidin	9.	niacin
2.	antioxidant	4.	inositol	6.	choline	8.	biotin	10.	folate

⑥ Chapter 11 – The Fat- Soluble Vitamins: ⑥ A, D, E, and K

Chapter Outline

I.   Vitamin A and Beta-Carotene
   A. Roles in the Body
      1. Vitamin A in Vision
      2. Vitamin A in Protein Synthesis and Cell Differentiation
      3. Vitamin A in Reproduction and Growth
      4. Beta-Carotene as an Antioxidant
   B. Vitamin A Deficiency
      1. Infectious Diseases
      2. Night Blindness
      3. Blindness (Xerophthalmia)
      4. Keratinization
   C. Vitamin A Toxicity
      1. Bone Defects
      2. Birth Defects
      3. Not for Acne
   D. Vitamin A Recommendations
   E. Vitamin A in Foods
      1. The Colors of Vitamin A Foods
      2. Vitamin A-Poor Fast Foods
      3. Vitamin A-Rich Liver
II. Vitamin D
   A. Roles in the Body
      1. Vitamin D in Bone Growth
      2. Vitamin D in Other Roles
   B. Vitamin D Deficiency

      1. Rickets
      2. Osteomalacia
      3. Osteoporosis
      4. The Elderly
   C. Vitamin D Toxicity
   D. Vitamin D Recommendations and Sources
      1. Vitamin D in Foods
      2. Vitamin D from the Sun
III. Vitamin E
   A. Vitamin E as an Antioxidant
   B. Vitamin E Deficiency
   C. Vitamin E Toxicity
   D. Vitamin E Recommendations
   E. Vitamin E in Foods
IV. Vitamin K
   A. Roles in the Body
   B. Vitamin K Deficiency
   C. Vitamin K Toxicity
   D. Vitamin K Recommendations and Sources
V.   The Fat-Soluble Vitamins—In Summary
VI. Antioxidant Nutrients in Disease Prevention
   A. Free Radicals and Disease
   B. Defending against Free Radicals
   C. Defending against Cancer
   D. Defending against Heart Disease
   E. Foods, Supplements, or Both?

Summing Up

Vitamin A is a part of visual (1)_____ and is essential for vision. It is involved in maintaining mucous (2)_____, helps maintain the skin, and is essential for the remodeling of (3)_____ during growth or mending. It plays a part in cell membranes, in (4)_____ synthesis, and in reproduction. Deficiency of vitamin A causes (5)_____ blindness; disorders of the respiratory, urogenital, reproductive, and nervous systems; and abnormalities of bones and teeth. (6)_____ symptoms are caused by large excesses (ten times the recommended intake or more) taken over a prolonged period and result only from the (7)_____ vitamin (from (8)_____ or animal products such as liver)—not from the precursor (9)_____ and its relatives.

Vitamin D promotes intestinal (10)_____ of calcium, mobilization of calcium from (11)_____ stores, and retention of calcium by the (12)_____, and is therefore essential for the calcification of bones and teeth. Given reasonable exposure to the (13)_____, human beings can synthesize this vitamin in the (14)_____. Deficiency of vitamin D causes

129

rickets in children and (15)_____ in adults; excesses cause abnormally high blood calcium levels and result in deposition of calcium crystals in soft tissues, such as the kidneys and major blood vessels.

The most substantiated role of vitamin E in human beings is as an (16)_____ that protects vitamin A and the polyunsaturated fatty acids (PUFA) from destruction by oxygen. Vitamin E also protects the lungs against oxidizing air (17)_____. Only one vitamin E deficiency has been confirmed in human beings: erythrocyte (18)_____.

Vitamin K, the coagulation vitamin, promotes normal blood (19)_____; deficiency causes hemorrhagic disease. The vitamin is synthesized by intestinal (20)_____ and is available from foods such as green vegetables and milk. Deficiency is normally seen only in (21)_____, whose intestinal flora have not become established, in people taking (22)_____ or sulfa drugs, or in people whose fat absorption is impaired.

Chapter Study Questions

1. List the fat-soluble vitamins. What characteristics do they have in common? How do they differ from the water-soluble vitamins?

2. Summarize the roles of vitamin A and the symptoms of its deficiency.

3. What is meant by *vitamin precursors*? Name the precursors of vitamin A, and tell in what classes of foods they are located. Give examples of foods with high vitamin A activity.

4. How is vitamin D unique among the vitamins? What is its chief function? What are the richest sources of this vitamin?

5. Describe vitamin E's role as an antioxidant. What are the chief symptoms of vitamin E deficiency?

6. What is vitamin K's primary role in the body? What conditions may lead to vitamin K deficiency?

Chapter Glossary

- **acne:** a chronic inflammation of the skin's follicles and oil-producing glands, which leads to an accumulation of oils inside the ducts that surround hairs; usually associated with the maturation of young adults.
- **alpha-tocopherol:** the active vitamin E compound.
- **beta-carotene:** one of the carotenoids; an orange pigment and vitamin A precursor found in plants.
- **carotenoids:** pigments commonly found in plants and animals, some of which have vitamin A activity. The carotenoid with the greatest vitamin A activity is beta-carotene.
- **cell differentiation:** the process by which immature cells develop specific functions different from those of the original that are characteristic of their mature cell type.
- **chlorophyll:** the green pigment of plants, which absorbs light and transfers the energy to other molecules, thereby initiating photosynthesis.
- **cholecalciferol:** an animal version of vitamin D (also called *vitamin D₃*)
- **cornea:** the transparent membrane covering the outside of the eye.
- **epithelial cells:** cells on the surface of the skin and mucous membranes.
- **epithelial tissue:** the layer of the body that serves as a selective barrier between the body's interior and the environment (examples are the cornea of the eyes, the skin, the respiratory lining of the lungs, and the lining of the digestive tract).
- **ergocalciferol:** a plant version of vitamin D (also called *vitamin D₂*).
- **erythrocyte:** red blood cell.
- **erythrocyte hemolysis:** the breaking open of red blood cells (erythrocytes); a symptom of vitamin E–deficiency disease in human beings.
- **fibrocystic breast disease:** a harmless condition in which the breasts develop lumps, sometimes associated with caffeine consumption. In some, it responds to abstinence from caffeine; in others, it can be treated with vitamin E.
- **hemolytic anemia:** the condition of having too few red blood cells as a result of erythrocyte hemolysis.
- **hemophilia:** a hereditary disease in which the blood is unable to clot because it lacks the ability to synthesize certain clotting factors.
- **hemorrhagic disease:** a disease characterized by excessive bleeding.
- **hypercalcemia:** high blood calcium that may develop from a variety of disorders, including vitamin D toxicity. It does *not* develop from a high calcium intake.
- **intermittent claudication:** severe calf pain caused by inadequate blood supply. It occurs when walking and subsides during rest.
- **international units (IU):** an old measure of vitamin activity used before direct chemical analysis was possible. Most food and supplement labels report their vitamin A contents using international units.
- **keratin:** a water-insoluble protein; the normal protein of hair and nails. Keratin-producing cells may replace mucus-producing cells in vitamin A deficiency.
- **keratinization:** accumulation of keratin in a tissue; a sign of vitamin A deficiency.
- **keratomalacia:** softening of the cornea that leads to irreversible blindness; seen in severe vitamin A deficiency.
- **lysosomes:** sacs of degradative enzymes found within cells (e.g. osteoclasts).
- **menadione:** the synthetic form of vitamin K.
- **mucous membranes:** the membranes, composed of mucus-secreting cells, that line the surfaces of body tissues.
- **muscular dystrophy:** a hereditary disease in which the muscles gradually weaken. Its most debilitating effects arise in the lungs.
- **night blindness:** slow recovery of vision after flashes of bright light at night or an inability to see in dim light; an early symptom of vitamin A deficiency.
- **opsin:** the protein portion of the visual pigment molecule.

- **osteoblasts:** cells that build bone.
- **osteoclasts:** cells that destroy bone during growth.
- **osteomalacia:** a bone disease characterized by softening of the bones. Symptoms include bending of the spine and bowing of the legs. The disease occurs most often in adult women.
- **phylloquinone:** the natural form of vitamin K.
- **pigment:** a molecule capable of absorbing certain wavelengths of light so that it reflects only those that we perceive as a certain color.
- **preformed vitamin A:** dietary vitamin A in its active form.
- **retinal:** the form of vitamin A that is active in vision and is also an intermediate in the conversion of retinol to retinoic acid.
- **rachitic rosary:** common name for the beaded ribs that result from poorly formed attachments of these bones to the cartilage; a sign of rickets.
- **remodeling:** the dismantling and re-formation of a structure, in this case, bone.
- **retina:** the layer of light-sensitive nerve cells lining the back of the inside of the eye; consists of rods and cones.
- **retinoic acid:** the form of vitamin A that acts like a hormone, regulating cell differentiation, growth, and embryonic development.
- **retinoids:** chemically related compounds with biological activity similar to that of retinol; metabolites of retinol.
- **retinol:** the form of vitamin A that supports reproduction and is the major transport and storage form of the vitamin.
- **retinol activity equivalents (RAE):** a measure of vitamin A activity; the amount of retinol that the body will derive from a food containing preformed retinol or its precursor beta-carotene.
- **retinol-binding protein (RBP):** the specific protein responsible for transporting retinol.
- **rhodopsin:** a light-sensitive pigment of the retina; contains the retinal form of vitamin A and the protein opsin.
- **rickets:** the vitamin D–deficiency disease in children characterized by inadequate mineralization of bone (manifested in bowed legs or knock-knees, outward-bowed chest, and knobs on ribs). A rare type of rickets, not caused by vitamin D deficiency, is known as *vitamin D–refractory rickets*.
- **sterile:** free of microorganisms, such as bacteria.
- **teratogenic:** causing abnormal fetal development and birth defects.
- **tocopherol:** a general term for several chemically related compounds, one of which has vitamin E activity.
- **tocopherol equivalents (TE):** the unit of measure used for vitamin E prior to 2000; these measures overestimated the amount of alpha-tocopherol.
- **vitamin A:** all naturally occurring compounds with the biological activity of retinol, the alcohol form of vitamin A.
- **vitamin A activity:** a term referring to both the active forms of vitamin A and the precursor forms in foods without distinguishing between them.
- **xanthophylls:** pigments found in plants; responsible for the color changes seen in autumn leaves.
- **xerophthalmia:** progressive blindness caused by severe vitamin A deficiency.
- **xerosis:** abnormal drying of the skin and mucous membranes; a sign of vitamin A deficiency.

Sample Test Questions

Select the best answer for each question.

1. A characteristic of the fat-soluble vitamins is that excesses are stored:
   a. in the liver and fatty tissues.
   b. in the blood.
   c. in epithelial cells.
   d. in mucous membranes.

2. Retinal is the _____ form of vitamin A.
   a. alcohol
   b. aldehyde
   c. acid
   d. provitamin A carotenoid

3. Night vision is maintained by:
   a. vitamin A, which combines with opsin in the dark to regenerate rhodopsin.
   b. vitamin D, which combines with opsin in the dark to regenerate rhodopsin.
   c. vitamin A, which combines with rhodopsin in the dark to regenerate retinene and opsin.
   d. vitamin E, which protects the polyunsaturated fatty acids in the membranes of the rod cells.

4. Which of the following surfaces is (are) not lined with epithelial cells?
   a. bladder and urethra
   b. mouth, stomach, and intestines
   c. eyelids
   d. lungs
   e. bones

5. Which vitamin is needed for bones to grow in length?
   a. vitamin A
   b. vitamin B_1
   c. vitamin C
   d. vitamin D

6. The active forms of vitamin A in the body are:
   a. retinol, retinal, and retinoic acid.
   b. retinol, retinyl esters, and opsin.
   c. retinol, retinal, and carotenoids.
   d. all of the above

7. Retinol-binding protein:
   a. converts retinol to retinal.
   b. converts retinyl esters to retinol.
   c. cleaves beta-carotene.
   d. carries vitamin A from the liver to the blood.

8. Cell differentiation refers to:
   a. the action of mucous membranes.
   b. immature cells developing functions different from those of the original.
   c. the ability of cells to serve as selective barriers.
   d. dismantling of a structure.

9. What is a vitamin A precursor found in plants?
   a. orange pigment
   b. beta-carotene
   c. carotenoid
   d. none of the above

10. The selective barrier between the body's interior and the environment is:
    a. epithelial tissue.
    b. the retina.
    c. mucous membranes.
    d. tocopherol.

11. An early sign of vitamin A deficiency is:
    a. rickets.
    b. hemolytic anemia.
    c. night blindness.
    d. total blindness.

12. Which of the following is a symptom of a vitamin A deficiency?
    a. anemia
    b. fissuring at the corners of the mouth
    c. keratinization of the epithelial cells
    d. erythematous areas closely resembling sunburn appearing on the skin

13. Vitamin A toxicity:
    a. can pose a teratogenic risk.
    b. is impossible.
    c. is helpful for clearing up acne.
    d. none of the above

14. Among fruits and vegetables, the best sources of vitamin A are those that are:
    a. green or yellow, such as lettuce and corn.
    b. dark green or deep orange, such as broccoli and sweet potatoes.
    c. green, such as lettuce, peas and snap beans.
    d. brightly colored, such as tomatoes and lemons.

15. Ultraviolet rays from the sun allow what vitamin to be synthesized?
    a. vitamin A   c. vitamin C
    b. vitamin B$_{12}$ d. vitamin D

16. Which of the following statements is true about vitamin D?
    a. It must be provided in the diet. c. It can be made in the body.
    b. It is an inorganic compound.   d. It is not needed in adulthood.

17. The most important physiological function of vitamin D is:
    a. synthesis of red blood cells.
    b. promotion of calcium and phosphorus utilization.
    c. increased resistance to disease.
    d. prevention of night blindness.

18. The vitamin D-deficiency disease of children is:
    a. xerophthalmia.   c. follicular hyperkeratosis.
    b. night blindness.   d. rickets.

19. The major role of vitamin E in the body is to:
    a. aid in normal blood clotting.   c. aid in formation of normal epithelial tissue.
    b. act as an antioxidant.   d. aid in protein metabolism.

20. A symptom of vitamin E deficiency that has been demonstrated in human beings is:
    a. weak bones.   c. reproductive failure.
    b. muscle paralysis.   d. breakage of red blood cell membranes.

21. A condition of having too few red blood cells resulting from erythrocyte hemolysis is:
    a. muscular dystrophy.   c. hemorrhagic disease.
    b. fibrocystic breast disease.   d. hemolytic anemia.

22. A disease characterized by inability of the blood to synthesize clotting factors is:
    a. hemolytic anemia.   c. hemophilia.
    b. osteomalacia.   d. intermittent claudication.

23. Vitamin E supplements may be beneficial to:
    a. people with intermittent claudication. c. women with "hot flashes."
    b. reverse damage caused by atherosclerosis. d bladder cancer patients.

24. Vitamin E supplementation:
    a. can cure hereditary muscular dystrophy in humans.
    b. if excessive, may interfere with blood clotting.
    c. is recommended for most people.
    d. is safe.

25. Vitamin E need varies with a person's intake of:
    a. polyunsaturated fatty acids.   c. cholesterol.
    b. saturated fatty acids.   d. other fat-soluble vitamins.

26. Vitamin K is necessary for:
    a. normal vision.
    b. normal blood clotting.
    c. normal muscle growth.
    d. prevention of night blindness.

27. If vitamin K is lacking, what condition may develop?
    a. thrombosis
    b. hemophilia
    c. hemorrhagic disease
    d. all of the above

28. A vitamin K deficiency may occur from:
    a. reduced fat absorption.
    b. antibiotic use.
    c. lack of sunlight.
    d. a and b
    e. a and c

29. Some of our vitamin K requirement is met by:
    a. synthesis of the vitamin by intestinal bacteria.
    b. synthesis of the vitamin from sunlight.
    c. synthesis of the vitamin from carotene.
    d. fortification of milk.

30. Toxicity of vitamin K:
    a. may cause death.
    b. is not common.
    c. may reduce the effectiveness of anticoagulant drugs.
    d. a and c
    e. b and c

Short Answer Questions

1. Characteristics of fat-soluble vitamins are:

    a.   c.

    b.   d.

2. The names of the fat-soluble vitamins are:

    a.   c.

    b.   d.

Problem Solving

1. If a carrot has 7,930 IU of beta-carotene, approximately how many μg RAE does it provide? If a supplement contains 20,000 IU of beta-carotene, approximately how many μg RAE does it provide?

2. What percentage of the U.S. RDA for vitamin A does an adult female receive from 1 oz. of liver?

Crossword Puzzle

Complete this crossword puzzle by Mary A. Wyandt, Ph.D., CHES.

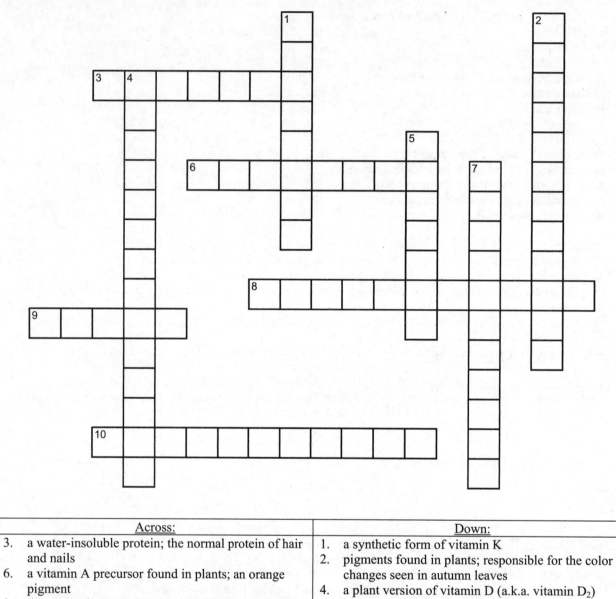

Across:		Down:	
3.	a water-insoluble protein; the normal protein of hair and nails	1.	a synthetic form of vitamin K
6.	a vitamin A precursor found in plants; an orange pigment	2.	pigments found in plants; responsible for the color changes seen in autumn leaves
8.	the green pigment of plants, which absorbs photons and transfers their energy to other molecules	4.	a plant version of vitamin D (a.k.a. vitamin D_2)
		5.	the alcohol form of vitamin A
9.	the protein portion of the visual pigment molecule	7.	red blood cell
10.	less active forms of vitamin E		

☙ Chapter 11 Answer Key ☙

Summing Up

1. pigments
2. membranes
3. bones
4. hormone
5. night
6. Toxicity
7. preformed
8. supplements
9. beta-carotene
10. absorption
11. bone
12. kidneys
13. sun
14. skin
15. osteomalacia
16. antioxidant
17. pollutants
18. hemolysis
19. clotting
20. bacteria
21. newborns
22. antibiotics

Chapter Study Questions

1. Vitamins A, D, E, and K. Found in the fat and oily parts of foods; stored primarily in the liver and adipose tissue. Fat-soluble vitamins are stored longer; therefore, daily intake is less crucial and toxicity risk is greater than for water-soluble vitamins.
2. Vitamin A is important in vision; maintenance of the cornea, epithelial cells, mucous membranes, and skin; bone and tooth growth; reproduction; hormone synthesis and regulation; immunity; and cancer protection. Symptoms of deficiency include anemia, diarrhea, general discomfort, depression, frequent respiratory, digestive, bladder, vaginal and other infections, abnormal tooth and jaw alignment, night blindness, keratinization, corneal degeneration leading to blindness, and rashes.
3. Compounds that can be converted into the active form of a vitamin. Carotenoids (the most active being beta-carotene) are available from plant foods; retinol compounds are available from animal foods. Examples of high-vitamin A activity foods are those with beta-carotene including spinach, winter squash, cantaloupe, carrots, sweet potatoes, and mango.
4. With sunlight, vitamin D can be made from cholesterol. Its chief function is to promote mineralization of bones. The richest sources are fortified milk, fortified margarine, eggs, liver, and fatty fish.
5. Vitamin E protects other substances from oxidation by being oxidized itself; it protects the lipids and other vulnerable components of the cell and its membranes from destruction. It is especially effective in preventing the oxidation of the polyunsaturated fatty acids. Deficiency symptoms include red blood cell breakage and anemia.
6. Synthesis of blood-clotting proteins and a blood protein that regulates blood calcium. Conditions which can lead to vitamin K deficiency are taking of antibiotics or sulfa drugs while consuming a diet low in vitamin K.

Sample Test Questions

1. a (p. 369)
2. b (p. 370)
3. a (p. 371)
4. e (p. 371)
5. a (p. 372)
6. a (p. 370)
7. d (p. 370)
8. b (p. 371)
9. b (p. 369)
10. a (p. 371)
11. c (p. 373)
12. c (p. 373)
13. a (p. 374)
14. b (p. 375)
15. d (p. 377)
16. c (p. 377)
17. b (p. 377)
18. d (p. 378)
19. b (p. 382)
20. d (p. 382)
21. d (p. 382)
22. c (p. 384)
23. a (p. 382)
24. b (p. 382)
25. a (p. 383)
26. b (p. 383)
27. c (p. 384)
28. d (p. 384)
29. a (p. 383)
30. e (p. 385)

Short Answer Questions

1. found in fat part of foods; stored primarily in liver and adipose tissue; need an average intake; toxicity is of concern
2. A, D, E, K

Problem Solving

1. 7,930 IU x 0.05 µg RAE/IU = 396.5 µg RAE
    20,000 IU x 0.15 µg RAE/IU = 3,000 µg RAE

2. 3 oz. liver, pan-fried = 6,582 µg RAE; RDA for adult female = 700 µg RAE
    6,582 µg RAE divided by 3 = 2,194 µg RAE
    2,194 divided by 700 x 100% = 313%

Crossword Puzzle

1. mendione
2. xanthophylls
3. keratin

4. ergocalciferol
5. retinol
6. carotene

7. erythrocyte
8. chlorophyll
9. opsin

10. tocotrienol

✪ Chapter 12 – Water and the Major Minerals ✪

Chapter Outline

I.   Water and the Body Fluids
   A.   Water Balance and Recommended Intakes
      1.   Water Intake
      2.   Water Sources
      3.   Water Losses
      4.   Water Recommendations
      5.   Health Effects of Water
   B.   Blood Volume and Blood Pressure
      1.   ADH and Water Retention
      2.   Renin and Sodium Retention
      3.   Angiotensin and Blood Vessel Constriction
      4.   Aldosterone and Sodium Retention
   C.   Fluid and Electrolyte Balance
      1.   Dissociation of Salt in Water
      2.   Electrolytes Attract Water
      3.   Water Follows Electrolytes
      4.   Proteins Regulate Flow of Fluids and Ions
      5.   Regulation of fluid and Electrolyte Balance
   D.   Fluid and Electrolyte Imbalance
      1.   Different Solutes Lost by Different Routes
      2.   Replacing Lost Fluids and Electrolytes
   E.   Acid-Base Balance
      1.   Regulation by the Buffers
      2.   Regulation in the Kidneys
II. The Minerals—An Overview
   A.   Inorganic Elements
   B.   The Body's Handling of Minerals
   C.   Variable Bioavailability
   D.   Nutrient Interactions
   E.   Varied Roles
III. Sodium
   A.   Sodium Roles in the Body
   B.   Sodium Recommendations
   C.   Sodium and Hypertension
   D.   Sodium and Bone Loss (Osteoporosis)
   E.   Sodium in Foods
   F.   Sodium Deficiency
   G.   Sodium Toxicity and Excessive Intakes
IV. Chloride
   A.   Chloride Roles in the Body

   B.   Chloride Recommendations and Intakes
   C.   Chloride Deficiency and Toxicity
V.   Potassium
   A.   Potassium Roles in the Body
   B.   Potassium Recommendations and Intakes
   C.   Potassium and Hypertension
   D.   Potassium Deficiency
   E.   Potassium Toxicity
VI. Calcium
   A.   Calcium Roles in the Body
      1.   Calcium in Bones
      2.   Calcium in Body Fluids
      3.   Calcium and Disease Prevention
      4.   Calcium and Obesity
      5.   Calcium Balance
      6.   Calcium Absorption
   B.   Calcium Recommendations and Sources
      1.   Calcium Recommendations
      2.   Calcium in Milk Products
      3.   Calcium in Other Foods
   C.   Calcium Deficiency
VII. Phosphorus
   A.   Phosphorus Roles in the Body
   B.   Phosphorus Recommendations and Intakes
VIII. Magnesium
   A.   Magnesium Roles in the Body
   B.   Magnesium Intakes
   C.   Magnesium Deficiency
   D.   Magnesium and Hypertension
   E.   Magnesium Toxicity
IX. Sulfate
X.   Osteoporosis and Calcium
   A.   Bone Development and Disintegration
   B.   Age and Bone Calcium
      1.   Maximizing Bone Mass
      2.   Minimizing Bone Loss
   C.   Gender and Hormones
   D.   Genetics and Ethnicity
   E.   Physical Activity and Body Weight
   F.   Smoking and Alcohol
   G.   Dietary Calcium
   H.   Other Nutrients
   I.   A Perspective on Supplements
   J.   Some Closing Thoughts

Summing Up

     Water in the body participates in many chemical (1)_____, and serves as a solvent,

(2)_____ medium, and lubricant. It makes up (3)_____ percent of the body's weight. The principal (4)_____ in body fluids are sodium, chloride, phosphorus, and (5)_____; each is maintained at a constant concentration by means of renal excretion. The electrolytes are involved in maintaining both water and acid-base (6)_____. (7)_____ and electrolyte imbalances are medical emergencies. Sodium is abundant in the diet, as part of (8)_____. Deficiencies are rare except in (9)_____. Potassium is an electrolyte that is important in (10)_____ balance. The best sources of potassium are fresh (11)_____ and (12)_____.

About 99 percent of the body's calcium is a structural component of the (13)_____ and teeth. The 1 percent of calcium found in body fluids helps maintain cell (14)_____ integrity, intercellular cohesion, transport of substances into and out of cells, and transmission of (15)_____ impulses. It is also essential for blood (16)_____ and acts as a cofactor in some enzyme systems. Calcium concentration in the blood is held constant.

Calcium deficiency may be caused directly by inadequate calcium intakes over a prolonged period or indirectly by vitamin (17)_____ deficiency. The diseases that result are rickets, osteomalacia, and (18)_____. A substantial portion of the recommended calcium intake is easily supplied by 2 cups of (19)_____ or equivalent dairy products such as cheese; (20)_____ soy milk is an alternative for people who are vegetarians, have milk (21)_____ or lactose (22)_____.

Phosphorus, another major mineral, is so abundant in foods that (23)_____ are unlikely. It participates with calcium in forming the (24)_____ of bone and therefore is found in large quantities in the body. Magnesium plays a role in the synthesis of body (25)_____ and so is important to all body functions. It is found lacking in human beings in conditions that aggravate dietary protein deficiency, such as kwashiorkor and (26)_____. A deficiency of magnesium causes (27)_____. Sulfur, like phosphorus, is a major mineral constituent of body (28)_____. It is (29)_____ in the diet, and deficiencies are unknown.

Chapter Study Questions

1. List the roles of water in the body.

2. List the sources of water intake and routes of water excretion.

3. What is ADH? Where does it exert its action? What is aldosterone? How does it work?

4. How does the body use electrolytes to regulate fluid balance?

5. What do the terms *major* and *trace* mean when describing the minerals in the body?

6. Describe some characteristics of minerals that distinguish them from vitamins.

7. What is the major function of sodium in the body? Describe how the kidneys regulate blood sodium. Is a dietary deficiency of sodium likely? Why or why not?

8. List calcium's roles in the body. How does the body keep blood calcium constant regardless of intake?

9. Name significant food sources of calcium. What are the consequences of inadequate intakes?

10. List the roles of phosphorus in the body. Discuss the relationship between calcium and phosphorus. Is a dietary deficiency of phosphorus likely? Why or why not?

11. State the major functions of chloride, potassium, magnesium, and sulfur in the body. Are deficiencies of these nutrients likely to occur in your own diet? Why or why not?

Chapter Glossary

- **adrenal glands:** glands adjacent to, and just above, each kidney.
- **aldosterone:** a hormone secreted by the adrenal glands that regulates blood pressure by increasing the reabsorption of sodium by the kidneys. Aldosterone also regulates chloride and potassium concentrations.
- **angiotensin:** a hormone involved in blood pressure regulation. Its precursor protein is called *angiotensinogen*; it is activated by *renin*, an enzyme from the kidneys.
- **anions:** negatively charged ions.
- **antidiuretic hormone (ADH):** a hormone produced by the pituitary gland in response to dehydration (or a high sodium concentration in the blood). It stimulates the kidneys to reabsorb more water and therefore to excrete less. It is also called *vasopressin*.
- **artesian water:** water drawn from a well that taps and confined aquifer in which the water is under pressure.
- **bicarbonate:** an alkaline compound with the formula HCO_3. It is produced in all cell fluids from the dissociation of carbonic acid to help maintain the body's acid-base balance. (Bicarbonate is also secreted from the pancreas during digestion as part of the pancreatic juice.)
- **binders:** chemical compounds in foods that combine with nutrients (especially minerals) to form complexes the body cannot absorb. Examples include phytates and oxalates.
- **bottled water:** drinking water sold in bottles.
- **calcitonin:** a hormone secreted by the thyroid gland that regulates blood calcium by lowering it when levels rise too high.
- **calcium:** the most abundant mineral in the body; found primarily in the body's bones and teeth.
- **calcium-binding protein:** a protein in the intestinal cells, made with the help of vitamin D, that facilitates calcium absorption.
- **calcium rigor:** hardness or stiffness of the muscles caused by high blood calcium concentrations.
- **calcium tetany:** intermittent spasm of the extremities due to nervous and muscular excitability caused by low blood calcium concentrations.
- **calmodulin:** an inactive protein that becomes active when bound to calcium. Once activated, it becomes a messenger that tells other proteins what to do. The system serves as an interpreter for hormone- and nerve-mediated messages arriving at cells.
- **carbonated water:** water that contains carbon dioxide gas, either naturally occurring or added, that causes bubbles to form in it; also called *bubbling* or *sparkling water*. Seltzer, soda, and tonic waters are legally soft drinks and are not regulated as water.
- **carbonic acid:** a compound with the formula H_2CO_3 that results from the combination of carbon dioxide (CO_2) and water (H_2O); of particular importance in maintaining the body's acid-base balance.
- **cations:** positively charged ions.
- **chloride:** the major anion in the extracellular fluids of the body. Chloride is the ionic form of chlorine, Cl^-; see Appendix B for a description of the chlorine-to-chloride conversion.
- **dehydration:** the condition in which body water output exceeds water input. Symptoms include thirst, dry skin and mucous membranes, rapid heartbeat, low blood pressure, and weakness.
- **dissociates:** physically separates.
- **distilled water:** water that has been vaporized and recondensed, leaving it free of dissolved minerals.
- **electrolyte solutions:** solutions that can conduct electricity.
- **electrolytes:** salts that dissolve in water and dissociate into charged particles called ions.
- **extracellular fluid:** fluid outside the cells. Extracellular fluid includes two main components—the interstitial fluid and plasma. Extracellular fluid accounts for approximately one-third of the body's water.

- **filtered water:** water treated by filtration, usually through *activated carbon filters* that reduce the lead in tap water, or by *reverse osmosis* units that force pressurized water across a membrane removing lead, arsenic, and some microorganisms from tap water.
- **fluid and electrolyte balance:** maintenance of the proper amounts and kinds of fluid and minerals in each compartment of the body fluids.
- **hard water:** water with a high calcium and magnesium content.
- **hydroxyapatite:** crystals made of calcium and phosphorus.
- **hypothalamus:** a brain center that controls activities such as maintenance of water balance, regulation of body temperature, and control of appetite.
- **interstitial fluid:** fluid between the cells (intercellular), usually high in sodium and chloride. Interstitial fluid is a large component of extracellular fluid.
- **intracellular fluid:** fluid within the cells, usually high in potassium and phosphate. Intracellular fluid accounts for approximately two-thirds of the body's water.
- **ions:** atoms or molecules that have gained or lost electrons and therefore have electrical charges. Examples include the positively charged sodium ion (Na^+) and the negatively charged chloride ion (Cl^-).
- **magnesium:** a cation within the body's cells, active in many enzyme systems.
- **major minerals:** essential mineral nutrients found in the human body in amounts larger than 5 g; sometimes called *macrominerals*.
- **milliequivalents (mEq):** the concentration of electrolytes in a volume of solution.
- Milliequivalents are a useful measure when considering ions because the number of charges reveals characteristics about the solution that are not evident when the concentration is expressed in terms of weight.
- **minerals:** inorganic elements.
- **mineral water:** water from a spring or well that typically contains 250 to 500 parts per million (ppm) of minerals. Minerals give water a distinctive flavor. Many mineral waters are high in sodium.
- **mineralization:** the process in which calcium, phosphorus, and other minerals crystallize on the collagen matrix of a growing bone, hardening the bone.
- **natural water:** water obtained from a spring or well that is certified to be safe and sanitary. The mineral content may not be changed, but the water may be treated in other ways such as with ozone or by filtration.
- **obligatory water excretion:** the amount of water the body has to excrete each day to dispose of its wastes (about 500 mL).
- **oral rehydration therapy (ORT):** a simple solution of sugar, salt, and water taken by mouth to treat dehydration caused by diarrhea.
- **osmosis:** the movement of water across a membrane toward the side where the solutes are more concentrated.
- **osmotic pressure:** the amount of pressure needed to prevent the movement of water across a membrane.
- **osteoporosis:** a disease in which the bones become porous and fragile due to a loss of minerals; also called adult bone loss.
- **parathyroid hormone:** a hormone from the parathyroid glands that regulates blood calcium by raising it when levels fall too low; also known as *parathormone*.
- **peak bone mass:** the highest attainable bone density for an individual, developed during the first three decades of life.
- **pH:** the unit of measure expressing a substance's acidity or alkalinity.
- **phosphorus:** a major mineral found mostly in the body's bones and teeth.
- **polar:** a neutral molecule that has opposite charges spatially separated within the molecule.
- **potassium:** the principal cation within the body's cells; critical to the maintenance of fluid balance, nerve impulse transmissions, and muscle contractions.
- **public water:** water from a municipal or county water system that has been treated and disinfected.
- **purified water:** water that has been treated by distillation or other physical or chemical processes that remove dissolved solids. Because purified water contains no minerals or contaminants, it is useful for medical and research purposes.
- **renin:** an enzyme from the kidneys that activates angiotensin.
- **salt:** a compound composed of a positive ion other than H^+ and a negative ion other than OH^-. An example is sodium chloride ($Na^+ Cl^-$).

- **salt sensitivity:** a characteristic of individuals who respond to a high salt intake with an increase in blood pressure or to a low salt intake with a decrease in blood pressure.
- **selectively permeable:** describes a membrane that allows the passage of some molecules, but not of others (e.g. cell membranes).
- **sodium:** the principal cation in the extracellular fluids of the body; critical to the maintenance of fluid balance, nerve impulse transmissions, and muscle contractions.
- **sodium-potassium pump:** a protein that regulates the flow of fluids and ions in and out of cells. The pump actively exchanges sodium for potassium across the cell membrane, using ATP as an energy source.
- **soft water:** water with a high sodium or potassium content.
- **solutes:** the substances that are dissolved in a solution. The number of molecules in a given volume of fluid is the solute concentration.
- **spring water:** water originating from an underground spring or well. It may be bubbly (carbonated), or "flat" or "still," meaning not carbonated. Brand names such as "Spring Pure" do not necessarily mean that the water comes from a spring.
- **sulfate:** the oxidized form of sulfur.
- **sulfur:** a mineral present in the body as part of some proteins.
- **thirst:** a conscious desire to drink.
- **vasoconstrictor:** a substance that constricts or narrows the blood vessels.
- **water balance:** the balance between water intake and output (losses).
- **water intoxication:** the rare condition in which body water contents are too high in all body fluid compartments.
- **well water:** water drawn from ground water by tapping into an aquifer.

Sample Test Questions

Select the best answer for each question.

1. The most essential nutrient is:
   a. carbohydrate.
   b. protein.
   c. fat.
   d. water.

2. The percentage of water in the body is about:
   a. 20%.
   c. 60%.
   b. 40%.
   d. 90%.

3. The proportion of water is smaller in a person who is:
   a. male.
   c. obese.
   b. muscular.
   d. active.

4. Water is involved in all of the following except:
   a. regulation of body temperature.
   c. shock absorption in the spinal cord.
   b. conversion of lipids to amino acids.
   d. lubrication of joints.

5. People must have water in their diets because the:
   a. kidneys exert no control over the amount of water excreted in the urine.
   b. fluid lost from the body must be replaced.
   c. kidneys must excrete a minimum amount of water in the urine to rid the body of wastes.
   d. a and b
   e. b and c

6. Fluid within the cells is called:
   a. intracellular fluid.
   c. extracellular fluid.
   b. interstitial fluid.
   d. ionic fluid.

144

7. The largest component of extracellular fluid is:
   a. intracellular fluid.
   b. interstitial fluid.
   c. electrolytes.
   d. ions.

8. Which of the following is false about dehydration?
   a. Water output exceeds water intake.
   b. Symptoms include thirst and dry skin.
   c. Symptoms include high blood pressure.
   d. Symptoms include rapid heartbeat and weakness.

9. The amount of water the body has to excrete each day to dispose of its wastes is:
   a. about 1500 mL.
   b. obligatory water excretion.
   c. insensible water losses.
   d. water intoxication.

10. The best water to drink for health reasons is:
    a. bottled water.
    b. carbonated water.
    c. soft water.
    d. hard water.

11. Which of the following substances is an electrolyte?
    a. water
    b. sodium
    c. a fatty acid
    d. glucose
    e. carbon

12. A cell membrane is an example of:
    a. a selectively permeable membrane.
    b. a milliequivalent.
    c. an anion.
    d. an electrolyte.

13. The amount and type of electrolytes in the body are largely regulated by:
    a. the liver.
    b. the kidneys.
    c. the intestines.
    d. the spleen.
    e. the pancreas.

14. Aldosterone secretion stimulates:
    a. sodium retention.
    b. sodium excretion.
    c. water excretion.
    d. a and c
    e. b and c

15. Buffers:
    a. move water into a concentrated solution.
    b. transport oxygen in the blood.
    c. maintain acid-base balance in the body.
    d. maintain blood pressure in the body.

16. Foods that have the highest sodium content include:
    a. fresh vegetables.
    b. foods that taste salty.
    c. meat products.
    d. processed foods.

17. The major anion of extracellular fluid is:
    a. sodium.
    b. calcium.
    c. potassium.
    d. sulfur.
    e. chloride.

18. People who respond to a high salt intake with an increase in blood pressure may have:
    a. dehydration.
    b. hypotension.
    c. salt cravings.
    d. salt sensitivity.

19. The *Dietary Guidelines* advise limiting daily salt intake to:
    a. 400 mg.
    b. 700 mg.
    c. 1500 mg.
    d. 2300 mg.

20. In cases of ultra-endurance competitive athletic events, athletes:
    a. should take salt tablets.
    b. may develop hyponatremia.
    c. are at risk for hypertension.
    d. should consume no more than 400 mg sodium per day.

21. The most reliable food source(s) of chloride is/are:
    a. meats and whole-grain cereals.
    b. salts and processed foods.
    c. dark green vegetables.
    d. public water.
    e. milk and milk products.

22. The richest food sources of potassium are:
    a. processed foods.
    b. ready-to-eat cereals.
    c. fresh foods of all kinds.
    d. cured meats.

23. Blood calcium levels can be increased by:
    1. better absorption of calcium in the intestines.
    2. retention of calcium by the kidneys.
    3. release of calcium from the bones.
    4. manufacture of calcium from albumin.
    5. release of calcium from the teeth.

    a. 1, 2
    b. 1, 2, 3
    c. 1, 2, 3, 5
    d. 3, 4

24. Factors that impair calcium balance include:
    a. vitamin D.
    b. a high-fiber diet.
    c. phytates.
    d. a and b
    e. b and c

25. An inadequate intake of calcium for many years can lead to:
    a. osteoporosis.
    b. impaired energy-nutrient metabolism.
    c. pernicious anemia.
    d. bleeding gums and ease in bruising.
    e. peptic ulcers.

26. A class of foods that is rich in calcium and phosphorus is:
    a. citrus fruits.
    b. milk and milk products.
    c. potato.
    d. carrots.
    e. liver and other organ meats.

27. Positive calcium balance is favored by:
    a. high fat intake.
    b. fiber.
    c. vitamin D.
    d. a and b

28. Phosphorus deficiencies are:
    a. rare.
    b. common.
    c. irreversible.
    d. a and b
    e. b and c

29. Magnesium:
    a. is directly necessary for protein synthesis in cells.
    b. protects bone structures against degeneration.
    c. is the body's principal intracellular electrolyte.
    d. is necessary for wound healing.
    e. helps maintain gastric acidity.

30. Sulfur:
    a. is found in the adipose tissue.
    b. is commonly deficient.
    c. has an RDA of 400 mg.
    d. is present in rigid proteins.

Short Answer Questions

1. Name the major minerals.

    a.   e.

    b.   f.

    c.   g.

    d.

2. Three sources of water for the body are:

    a.

    b.

    c.

3. Four routes of water excretion are:

    a.

    b.

    c.

    d.

Problem Solving

1. The soup recipe calls for 2 tsp of salt. How many grams of salt is that? How many grams of sodium are in that much salt?

2. If a cup of milk provides 300 mg of calcium and you are trying to consume 800 mg of calcium per day, how many cups of milk do you need to drink (assuming this is your only source of calcium)?

Crossword Puzzle

Complete this crossword puzzle by Mary A. Wyandt, Ph.D., CHES.

Across:		Down:	
2.	water that has been treated by distillation or other physical or chemical processes	1.	water from a spring or well that typically contains 250-500 ppm of minerals
7.	water that has been vaporized and recondensed, leaving it free of dissolved minerals	3.	water drawn from ground water by tapping in an aquifer
8.	water originating from an underground spring or well	4.	water treated by filtration, usually through active carbon filters that reduce lead in tap water
9.	water from a municipal or county water system	5.	water with a high calcium and magnesium content
10.	water obtained from a spring or well that is certified to be safe and sanitary	6.	water with a high sodium or potassium content

✆ Chapter 12 Answer Key ✆

Summing Up

1. reactions
2. transportation
3. 55 to 60
4. electrolytes
5. potassium
6. balance
7. Fluid
8. salt
9. dehydration
10. fluid
11. fruits
12. vegetables
13. bones
14. membrane
15. nerve
16. clotting
17. D
18. osteoporosis
19. milk
20. fortified
21. allergy
22. intolerance
23. deficiencies
24. crystals
25. proteins
26. alcoholism
27. tetany
28. tissues
29. abundant

Chapter Study Questions

1. Water carries nutrients and waste products throughout the body; helps to form the structure of macromolecules; actively participates in chemical reactions; fills the cells and the spaces between them; serves as the solvent for minerals, vitamins, amino acids, glucose, and many other small molecules; acts as a lubricant around joints; serves as shock absorber inside the eyes, spinal cord, and the amniotic sac during pregnancy; and aids in the body's temperature regulation.
2. Sources of intake: liquids, foods, metabolic water. Routes of excretion: kidneys, skin, lungs, feces.
3. ADH is a hormone released by the pituitary gland in response to highly concentrated blood. It exerts its action on the kidneys, and they respond by reabsorbing water, thus preventing water loss. Aldosterone is a hormone secreted by the adrenal glands that stimulates the reabsorption of sodium by the kidneys.
4. The body uses electrolytes to control the movement of water inside cells and between cells. Since water is attracted to electrolytes and will follow them, the cells can induce water to flow across a membrane by moving electrolytes to the other side. Cells may use transport proteins to move electrolyte ions across the cell membrane.
5. Major minerals are needed in the largest amounts in the body; trace minerals are needed in small amounts in the body. Both are equally important.
6. Minerals can become charged particles and can form compounds; minerals retain their chemical identity because they are inorganic elements rather than organic molecules made up of elements.
7. Na's major function is to maintain fluid and electrolyte balance. The kidneys retain or excrete sodium and water to regulate blood pressure. A dietary deficiency of sodium is not likely because of sodium's abundance in food products.
8. Ca's roles: bone structure, cell membrane integrity, transport of ions, muscle action, nerve impulses, regulates blood vessel wall muscle tone, helps regulate blood pressure, aids blood clotting, acts as cofactor for enzymes. When levels fall, intestinal absorption increases, bone withdrawal increases, and kidney excretion diminishes; these processes are regulated by a system of hormones and vitamin D. Excessive intakes of calcium can cause kidney stone formation.
9. Calcium is found predominantly in milk and milk products. Inadequate intakes limit the bones' ability to achieve optimal mass and density and increase risk of osteoporosis and associated fractures.
10. P's roles: bone structure, part of DNA and RNA, activates enzymes, part of ATP, part of lipid structure and cell membranes. They combine as calcium phosphate in the crystals of bone and teeth, providing strength and rigidity. A deficiency is unlikely because phosphorus is abundant in animal tissues, eggs, and milk.
11. Chloride: maintains pH balance, allows nerve transmission and muscle contraction, catalyst in metabolism; potassium: maintains fluid and electrolyte balance and cell integrity; magnesium: bone structure, protein synthesis, energy metabolism, muscle relaxation; sulfur: protein structure. Deficiency of chloride is unusual but it can cause fluid and electrolyte problems; deficiency of potassium can cause increase in blood pressure, kidney stones and bone turnover.

Sample Test Questions

1. d (p. 397)
2. c (p. 397)
3. c (p. 397)
4. b (p. 397)
5. e (p. 399)
6. a (p. 398)
7. b (p. 398)
8. c (p. 398)
9. b (p. 399)
10. d (p. 400)
11. b (p. 403)
12. a (p. 404-405)
13. b (p. 406)
14. a (p. 401)
15. c (p. 407)
16. d (p. 411)
17. e (p. 413)
18. d(p. 410)
19. d (p. 411)
20. b (p. 412)
21. b (p. 413)
22. c (p. 414)
23. b (p. 416-417)
24. e (p. 418)
25. a (p. 421)
26. b (p. 418)
27. c (p. 418)
28. a (p. 422)
29. a (p. 424)
30. d (p. 427)

Short Answer Questions

1. calcium, phosphorus, potassium, sulfur, sodium, chloride, magnesium
2. liquids, foods, metabolic water
3. kidneys, lungs, feces, skin

Problem Solving

1. 2 tsp x 5 g/tsp = 10 g salt
   2 tsp salt x 2 g/tsp salt = 4 g sodium   OR   10 g salt x 400 mg/g salt = 4000 mg = 4 g salt

2. 800 mg divided by 300 mg/c = 2.7 c (rounded)

Crossword Puzzle

1. mineral water
2. purified water
3. well water
4. filtered water
5. hard water
6. soft water
7. distilled water
8. spring water
9. public water
10. natural water

✆ Chapter 13 – The Trace Minerals ✆

Chapter Outline

Summing Up

Iron is found principally in the red blood cells, where it comprises part of the (1)_____-

carrier protein hemoglobin. When red blood cells die and are dismantled in the (2)_____, the iron is retrieved and transported by iron-carrier proteins back to bone (3)_____, where new red blood cells are synthesized. There is no route of excretion for iron; losses are small except when (4)_____ is lost, as in (5)_____ or hemorrhage. Thus women's needs for iron are ordinarily (6)_____ than men's.

Iron-deficiency (7)_____, one of the world's most widespread malnutrition problems, is most common in women and (8)_____. Furthermore, (9)_____ iron deficiency is a vast world malnutrition problem, limiting people's working ability. Food sources of iron for women must be chosen carefully if even two-thirds of the recommended intake is to be met within a (10)_____ allowance that is not excessive. The (11)_____ of foods somewhat improves women's iron intakes, but may produce iron (12)_____ in men. Addition of an iron (13)_____ to the diet may be advisable for some women.

(14)_____, fish, and poultry are superior sources of iron. (15)_____ or enriched breads and cereals, legumes, dark greens, and some fruits are other good sources of iron. Foods in the (16)_____ group are notable for their lack of iron. Iron absorption is enhanced by vitamin (17)_____, other organic acids, sugars, and the (18)_____ factor; it is reduced by phytates, fibers, soy, legumes, (19)_____, and coffee. Zinc appears in every body tissue and supports several physiological functions, including normal growth and (20)_____ development. As with iron, a homeostatic mechanism may regulate the amount of zinc (21)_____ by the body.

(22)_____ of zinc have been observed in some children in the United States and Canada as well as in the Middle East. The richest food sources of zinc include (23)_____, meat, and liver. Milk, eggs, and whole-grain products are also good sources. When estimating the amount of zinc in a diet, one must take into account all of the factors which affect its (24)_____.

Iodide forms part of the (25)_____ hormones; deficiency may cause simple (26)_____, slowed metabolism, and cretinism. The use of iodized (27)_____ protects against deficiency. Selenium acts as cofactor for an (28)_____ enzyme. A severe deficiency can cause (29)_____ failure. Selenium-poor soil may correlate with certain kinds of (30)_____. Copper is important for red blood cell formation, (31)_____ synthesis, and central nervous system function.

Manganese aids many body (32)_____, but its safe range is narrow, with toxicity causing a severe (33)_____-disease syndrome in human beings. The (34)_____ ion combines with calcium and phosphorus to stabilize the (35)_____ structure of bones and teeth. In communities where the water contains fluoride, dental (36)_____ are less prevalent than in communities where the water supply is low in fluoride. Chromium, as part of the (37)_____ tolerance factor, works with (38)_____ in promoting glucose uptake into cells and normal carbohydrate metabolism. Chromium deficiency is believed to be responsible for some cases of adult-onset (39)_____ and to cause growth failure in children with protein-energy malnutrition.

Molybdenum functions as a part of several (40)_____ systems. Other (41)_____ minerals with known physiological roles include nickel, silicon, tin, vanadium, and cobalt.

Chapter Study Questions

1. Distinguish between heme and nonheme iron. Discuss the factors that enhance iron absorption.

2. Distinguish between iron deficiency and iron-deficiency anemia. What are the symptoms of iron-deficiency anemia?

3. What causes iron overload? What are its symptoms?

4. Describe the similarities and differences in the absorption and regulation of iron and zinc.

5. Discuss possible reasons for a low intake of zinc. What factors affect the bioavailability of zinc?

6. Describe the principal functions of iodide, selenium, copper, manganese, fluoride, chromium and molybdenum in the body.

7. What public health measure has been used in preventing simple goiter? What measure has been recommended for protection against tooth decay?

8. Discuss the importance of balanced and varied diets in obtaining the essential minerals and avoiding toxicities.

9. Describe some of the ways trace minerals interact with each other and with other nutrients.

Chapter Glossary

- **chelate:** a substance that can grasp the positive ions of a metal.
- **cofactor:** a substance that works with an enzyme to facilitate a chemical reaction.
- **contamination iron:** iron found in foods as the result of contamination by inorganic iron salts from iron cookware, iron-containing soils, and the like.
- **cretinism:** a congenital disease characterized by mental and physical retardation and commonly caused by maternal iodine deficiency during pregnancy.
- **enteropancreatic circulation:** the circulatory route from the pancreas to the intestine and back to the pancreas.
- **erythrocyte protoporphyrin:** a precursor to hemoglobin.
- **ferric iron:** iron in its oxidized ionic state; Fe^{+++}.
- **ferritin:** the iron storage protein.
- **ferrous iron:** iron in its reduced ionic state; Fe^{++}.
- **fluorapatite:** the stabilized form of bone and tooth crystal, in which fluoride has replaced the hydroxyl groups of hydroxyapatite.
- **fluorosis:** discoloration and pitting of tooth enamel caused by excess fluoride during tooth development.
- **molybdenum:** a trace element.
- **geophagia:** clay eating (a form of pica).
- **glucose tolerance factors (GTF):** small organic compounds that enhance insulin's action. Some glucose tolerance factors contain chromium.
- **goiter:** an enlargement of the thyroid gland due to an iodine deficiency, malfunction of the gland, or overconsumption of a goitrogen.
- **goitrogen:** a substance that enlarges the thyroid gland and causes *toxic goiter*. Goitrogens occur naturally in such foods as cabbage, kale, brussels sprouts, cauliflower, broccoli, and kohlrabi.
- **heavy metal:** any of a number of mineral ions such as mercury and lead, so called because they are of relatively high atomic weight. Many heavy metals are poisonous.
- **hematocrit:** measurement of the volume of the red blood cells packed by centrifuge in a given volume of blood.
- **heme:** the iron-holding part of the hemoglobin and myoglobin proteins. About 40% of the iron in meat, fish, and poultry is bound into heme; the other 60% is *nonheme* iron.
- **hemochromatosis:** a genetically determined failure to prevent absorption of unneeded dietary iron that is characterized by iron overload and tissue damage.
- **hemoglobin:** the oxygen-carrying protein of the red blood cells that transports oxygen from the lungs to tissues throughout the body; hemoglobin accounts for 80% of the body's iron.
- **hemosiderin:** an iron-storage protein primarily made in times of iron overload.
- **hemosiderosis:** a condition characterized by the deposition of hemosiderin in the liver and other tissues.
- **hepcidin:** a hormone produced by the liver that regulates iron balance.
- **iron deficiency:** the state of having depleted iron stores.
- **iron-deficiency anemia:** severe depletion of iron stores that results in low hemoglobin and small, pale red blood cells. Anemias that impair hemoglobin synthesis are *microcytic* (small cell).
- **iron overload:** toxicity from excess iron.

- **Keshan disease:** the heart disease associated with selenium deficiency, named for one of the provinces of China where it was studied. Keshan disease is characterized by heart enlargement and insufficiency; fibrous tissue replaces the muscle tissue that normally composes the middle layer of the walls of the heart.
- **metalloenzymes:** enzymes that contain one or more minerals as part of their structures.
- **metallothionein:** a sulfur-rich protein that avidly binds with and transports metals such as zinc.
- **MFP factor:** a peptide released during the digestion of meat, fish, and poultry that enhances nonheme iron absorption.
- **microcytic hypochromic anemia:** anemia characterized by small, pale red blood cells (e.g., iron-deficiency anemia).
- **mucosa:** a mucous membrane such as the one that lines the GI tract. The adjective of mucosa is *mucosal*.
- **myoglobin:** the oxygen-holding protein of the muscle cells.
- **pagophagia:** ice craving (a form of pica).
- **pica:** a craving for nonfood substances.
- **selenium:** a trace element.
- **simple goiter:** goiter caused by iodine deficiency.
- **thyroxine:** common name for tetraiodothyronine (T_4), a hormone released by the thyroid gland to its target tissues. Upon reaching the cells, T_4 is deiodinated to triiodothyronine (T_3), which is the active form of the hormone.
- **trace minerals:** essential mineral nutrients found in the human body in amounts smaller than 5 g; sometimes called microminerals.
- **transferrin:** the iron transport protein.

Sample Test Questions

Select the best answer for each question.

1. Essential mineral nutrients found in the human body in amounts smaller than 5 grams are called:
   a. macrominerals.
   b. major minerals.
   c. trace minerals.
   d. minor minerals.

2. Which of the following is true about microminerals?
   a. The body requires them in large amounts.
   b. They participate in specific tasks only in one area of the body.
   c. The micromineral contents of foods are unaffected by soil composition.
   d. Factors in the diet and within the body affect the bioavailability of minerals.

3. The most common result of a trace mineral deficiency in children is:
   a. GI tract problems.
   b. failure to grow and thrive.
   c. problems with muscles.
   d. central nervous system failure.

4. Since toxicities can easily develop at intakes of trace minerals not far above the requirements,
   a. the supplement industry prepares products below the recommendations.
   b. the FDA requires that amounts of trace minerals in supplements be limited.
   c. consumers must be aware of possible dangers and select supplements carefully.
   d. a and b
   e. a and c

5. Which of the following statements is true about interactions among the trace minerals?
   a. Interactions are uncommon and unnecessary.
   b. Interactions may lead to nutrient imbalances.
   c. Taking high doses of supplements benefits interactions of trace minerals.
   d. a and b
   e. b and c

6. Iron is important in the body because it is:
   a. needed for blood clotting.
   b. an integral part of bones and teeth.
   c. a constituent of hemoglobin.
   d. an antioxidant.

7. Which of the following iron-containing compounds transports iron from the intestine to the liver and bone marrow?
   a. ferritin
   b. myoglobin
   c. transferrin
   d. hemoglobin

8. The most common symptom of an iron deficiency is:
   a. night blindness.
   b. abnormal blood clotting.
   c. anemia.
   d. a skin rash.
   e. hemophilia.

9. Pica is:
   a. an appetite for ice, clay, paste or other nonfood substances.
   b. a mineral deficiency associated with a certain area of the country.
   c. a localized skin rash.
   d. a cancer-causing additive.

10. The hemoglobin level in the blood is used to assess a person's _____ status.
    a. copper
    b. folate
    c. vitamin B_{12}
    d. iron
    e. zinc

11. The iron-storage protein made in times of iron overload is:
    a. myoglobin.
    b. hemosiderin.
    c. heme.
    d. transferrin.

12. The measurement of the volume of red blood cells packed in a given volume of blood is:
    a. ferritin.
    b. hepcidin.
    c. hematocrit.
    d. heme.

13. Which of the following is characterized by iron overload and tissue damage?
    a. hemochromatosis
    b. contamination iron
    c. erythrocyte protoporphyrin
    d. hemosiderosis

14. People at high risk for iron deficiency include:
    a. older men and women.
    b. middle-aged men.
    c. infants and young children.
    d. all of the above

15. Which food group supplies most of the iron in the American diet?
    a. meat and beans group
    b. milk group
    c. vegetable group
    d. grains group

16. A very poor source of iron is:
    a. legumes.
    b. dried fruits.
    c. whole-grain breads.
    d. milk.

17. An important function of zinc is to serve as a(n):
    a. enzyme.
    b. cofactor.
    c. protein carrier.
    d. oxygen carrier.

18. Zinc is involved in the circulatory route from the pancreas to the intestine and back to the pancreas. This is called:
    a. phytate enhancement.
    b. oxalate-binding cofactor.
    c. ligand-operating absorption.
    d. cofactor assistance.
    e. enteropancreatic circulation.

19. Zinc deficiency symptoms include all but which of the following?
    a. severe growth retardation
    b. arrested sexual maturation
    c. decreased taste sensitivity
    d. pernicious anemia

20. Among the following, the best sources of available zinc are:
    a. shellfish, meats, and liver.
    b. breads, cereals, and grains.
    c. fruits and vegetables.
    d. milk products.

21. Which of the following statements is true about zinc supplementation?
    a. People in developed countries should take zinc supplements.
    b. Zinc lozenges are effective at reducing incidence of the common cold.
    c. Zinc lozenges have no side effects.
    d. In developing countries, zinc supplements play a major role in treatment of childhood infectious diseases.

22. Consumption of which of the following would best help to insure normal iodide intake?
    a. liver
    b. irradiated milk
    c. salt-water fish
    d. fortified margarine

23. Cretinism is caused by a deficiency of:
    a. copper.
    b. iron.
    c. fluoride.
    d. zinc.
    e. iodide.

24. One of copper's roles is to:
    a. assist in consumption of oxygen and oxygen radicals.
    b. manufacture collagen.
    c. assist in oxidation of ferrous iron to ferric iron.
    d. a and b
    e. a, b, and c

25. Which of the following statements is true about manganese?
    a. Manganese deficiencies are common in children.
    b. Manganese requirements are high.
    c. High intakes of calcium and iron limit manganese absorption.
    d. Manganese is found most in meat products.

26. Fluoride is necessary nutritionally for:
    a. hardness of the bones and teeth.
    b. production of the thyroid hormone.
    c. prevention of anemia.
    d. the metabolism of glucose.
    e. a and b

27. Excess fluoride can cause:
    a. deposition of calcium in soft tissues.
    b. a drastic increase in tooth decay.
    c. fluorosis.
    d. osteomalacia.

28. Selenium is involved in:
    a. hormone production.
    b. protein synthesis.
    c. antioxidant activities.
    d. glycogen breakdown.

29. Contaminant minerals impair the body's growth and work capacity. They:
    a. enter the food supply by way of soil, water and air pollution.
    b. include lead and mercury.
    c. are food additives.
    d. a and b
    e. a, b and c

30. Cobalt seems to be important in nutrition as part of:
    a. vitamin A.
    b. glucose tolerance factor.
    c. vitamin B_{12}.
    d. vitamin B_6.

Short Answer Questions

1. Name the trace minerals.

    a.   f.   k.

    b.   g.   l.

    c.   h.   m.

    d.   i.   n.

    e.   j.

2. Iron's two ionic states are:

    a.   b.

Crossword Puzzle

Complete this crossword puzzle by Mary A. Wyandt, Ph.D., CHES.

Across:	Down:
3. a condition characterized by the deposition of hemosiderin in the liver and other tissues	1. a congenital disease characterized by mental and physical retardation and commonly caused by iodine deficiency during pregnancy
5. a trace element (symbol: Mo)	2. the iron-holding part of the hemoglobin and myoglobin proteins
7. the oxygen-carrying protein of the red blood cells that transport oxygen from the lungs to tissues throughout the body	4. a trace element (symbol: Se)
8. measurement of the volume of the red blood cells packed by centrifuge in a given volume of blood	5. the oxygen-holding protein of the muscle cells
9. the stabilized form of bone and tooth crystal	6. an enlargement of the thyroid gland due to iodine deficiency

✆ Chapter 13 Answer Key ✆

Summing Up

1. oxygen
2. liver
3. marrow
4. blood
5. menstruation
6. greater
7. anemia
8. children
9. marginal
10. kcalorie
11. enrichment
12. overload
13. supplement
14. Meats
15. Whole-grain
16. milk
17. C
18. MFP
19. tea
20. sexual
21. absorbed
22. Deficiencies
23. shellfish
24. availability
25. thyroid
26. goiter
27. salt
28. antioxidant
29. heart
30. cancers
31. collagen
32. enzymes
33. brain
34. fluoride
35. crystalline
36. caries
37. glucose
38. insulin
39. diabetes
40. enzyme
41. trace

Chapter Study Questions

1. Heme iron is contained in the organic molecule heme and is better absorbed than non-heme iron. Vitamin C, organic acids, sugars, and MFP factor enhance iron absorption.
2. Iron deficiency is the state of being without iron stores; iron-deficiency anemia is a condition of small and pale blood cells resulting from an iron deficiency that causes pallor, fatigue, weakness, headaches, and apathy.
3. Iron overload is usually caused by a gene that enhances iron absorption. Symptoms: fatigue, headache, irritability, lowered work performance, anemia.
4. Similarities exist in absorption and regulation: absorption of both is partially determined by a person's status; other dietary factors can inhibit or facilitate absorption. Differences exist in that the body receives zinc from both ingested food and zinc-rich pancreatic secretions; the storage methods for each vary.
5. Diets very low in animal protein may be inadequate in zinc. Phytates and fiber.
6. Iodide: part of thyroid hormones; selenium: part of an enzyme that acts as an antioxidant; copper: part of enzymes, catalyst in hemoglobin formation, collagen synthesis and wound healing, maintains nerve fiber sheaths; manganese: acts as a cofactor for many enzymes that facilitate the metabolism of carbohydrate, lipids, and amino acids; fluoride: part of crystal structure of bones and teeth; chromium: part of glucose tolerance factor; molybdenum: acts as a working part of several metalloenzymes.
7. Iodization of salt. Fluoridation of water.
8. Such a diet prevents an excess of one trace mineral from causing a deficiency of another, or of a deficiency which may cause a toxic reaction of another, provides factors that promote trace mineral absorption, and includes food sources that contain all the trace minerals.
9. Fiber and phytates bind zinc, limiting its bioavailability; large doses of iron inhibit zinc absorption; large doses of zinc inhibit iron and copper absorption.

Sample Test Questions

1. c (p. 441)
2. d (p. 441)
3. b (p. 441)
4. c (p. 441-442)
5. b (p. 442)
6. c (p. 443)
7. c (p. 443)
8. c (p. 445-446)
9. a (p. 447)
10. d (p. 446)
11. b (p. 445, 448)
12. c (p. 446)
13. a (p. 448)
14. c (p. 446)
15. a (p. 449)
16. d (p. 449)
17. b (p. 452)
18. e (p. 452)
19. d (p. 453-454)
20. a (p. 454)
21. d (p. 455)
22. c (p. 456)
23. e (p. 456)
24. e (p. 458)
25. c (p. 459)
26. a (p. 460)
27. c (p. 460)
28. c (p. 457)
29. d (p. 463)
30. c (p. 462)

Short Answer Questions

1. iron, zinc, iodide, copper, manganese, fluoride, chromium, selenium, molybdenum, nickel, silicon, tin, cobalt, arsenic
2. ferrous (reduced, Fe^{++}); ferric (oxidized, Fe^{+++})

Crossword Puzzle

1. cretinism
2. heme
3. hemosiderosis

4. selenium
5A. molybdenum
5D. myoglobin

6. goiter
7. hemoglobin
8. hematocrit

9. fluorapatite

☙ Chapter 14 – Fitness: ❧
Physical Activity, Nutrients, and Body Adaptations

Chapter Outline

I. Fitness
  A. Benefits of Fitness
  B. Developing Fitness
    1. The Overload Principle
    2. The Body's Response to Physical Activity
    3. Cautions on Starting
  C. Cardiorespiratory Endurance
    1. Cardiorespiratory Conditioning
    2. Muscle Conditioning
    3. A Balanced Fitness Program
  D. Weight Training
II. Energy Systems, Fuels, and Nutrients to Support Activity
  A. The Energy Systems of Physical Activity—ATP and CP
    1. ATP
    2. CP
    3. The Energy-Yielding Nutrients
  B. Glucose Use during Physical Activity
    1. Diet Affects Glycogen Storage and Use
    2. Intensity of Activity Affects Glycogen Use
    3. Lactic Acid
    4. Duration of Activity Affects Glycogen Use
    5. Glucose Depletion
    6. Glucose during Activity
    7. Glucose after Activity
    8. Training Affects Glycogen Use
  C. Fat Use during Physical Activity
    1. Duration of Activity Affects Fat Use
    2. Intensity of Activity Affects Fat Use
    3. Training Affects Fat Use
  D. Protein Use during Physical Activity—and Between Times
    1. Protein Used in Muscle Building
    2. Protein Used as Fuel
    3. Diet Affects Protein Use during Activity
    4. Intensity and Duration of Activity Affect Protein Use during Activity
    5. Training Affects Protein Use

    6. Protein Recommendations for Active People
  E. Vitamins and Minerals to Support Activity
    1. Supplements
    2. Vitamin E
    3. Iron Deficiency
    4. Iron-Deficiency Anemia
    5. Sports Anemia
    6. Iron Recommendations for Athletes
  F. Fluids and Electrolytes to Support Activity
    1. Fluid Losses via Sweat
    2. Hyperthermia
    3. Hypothermia
    4. Fluid Replacement via Hydration
    5. Electrolyte Losses and Replacement
    6. Hyponatremia
  G. Poor Beverage Choices: Caffeine and Alcohol
    1. Caffeine
    2. Alcohol
III. Diets for Physically Active People
  A. Choosing a Diet to Support Fitness
    1. Water
    2. Nutrient Density
    3. Carbohydrate
    4. Protein
    5. A Performance Diet Example
  B. Meals Before and After Competition
    1. Pregame Meals
    2. Postgame Meals
IV. Supplements as Performance-Enhancing Aids
  A. Ergogenic Aids
  B. Dietary Supplements
    1. Carnitine
    2. Chromium Picolinate
    3. Complete Nutrition Supplements
    4. Creatine
    5. Conjugated Linoleic Acid
    6. Caffeine
    7. Oxygenated Water
  C. Hormonal Supplements
    1. Anabolic Steroids
    2. DHEA and Androstenedione
    3. Human Growth Hormone

Summing Up

In the body, nutrition and (1)_____ go hand in hand. The working body demands that the energy-yielding nutrients provide (2)_____ for the increased metabolism of exercise. Lack of exercise can lead to cardiovascular disease, (3)_____, intestinal disorders, apathy, (4)_____, and accelerated (5)_____ losses. The four components of fitness are flexibility, (6)_____, muscle endurance, and (7)_____ endurance.

Exercises that improve cardiovascular endurance raise the heart rate for more than (8)_____ minutes and use most of the large muscle groups of the body. Exercises are classified as (9)_____ and anaerobic.

The first fuels of exercise are adenosine triphosphate (ATP) and (10)_____ (CP). Glucose, stored as (11)_____, is essential for performance. A high- (12)_____ diet is most beneficial for athletic performance. Intensity and (13)_____ of exercise affect use of exercise fuels.

Vitamins and minerals assist in releasing energy from fuels and transporting (14)_____. An adequate diet will usually supply the necessary vitamins and minerals for athletes and (15)_____ are generally not recommended. Oversupplementation may actually (16)_____ athletic performance.

Dehydration is a concern for athletes. Full hydration is imperative for athletes during (17)_____ as well as during competition. Plain water or diluted juice is recommended. (18)_____ loading is used to trick the muscles into storing extra glycogen before competition. This technique has been altered to minimize undesirable side effects.

A diet that provides ample (19)_____ and consists of a variety of nutrient-dense foods in quantities to meet (20)_____ needs will not only enhance athletic performance but overall health as well.

Chapter Study Questions

1. Define fitness, and list its benefits.

2. Explain the overload principle.

3. Define cardiorespiratory endurance and list some of its benefits.

4. What types of exercise are aerobic? Which are anaerobic?

5. Describe the relationships among energy expenditure, type of activity, and oxygen use.

6. What factors influence the body's use of glucose during physical activity? How?

7. What factors influence the body's use of fat during physical activity? How?

8. What factors influence the body's use of protein during physical activity? How?

9. Why are some athletes likely to develop iron-deficiency anemia? Compare iron-deficiency anemia and sports anemia, explaining the differences.

10. Discuss the importance of hydration during training, and list recommendations to maintain fluid balance.

11. Describe the components of a healthy diet for athletic performance.

Chapter Glossary

- **atrophy:** becoming smaller; with regard to muscles, a decrease in size (and strength) because of disuse, undernutrition, or wasting diseases.
- **body composition:** the proportions of muscle, bone, fat, and other tissue that make up a person's total body weight.
- **carbohydrate loading:** a regimen of moderate exercise followed by the consumption of a high-carbohydrate diet that enables muscles to store glycogen beyond their normal capacities; also called *glycogen loading* or *glycogen super compensation.*
- **cardiac output:** the volume of blood discharged by the heart each minute; determined by multiplying the stroke volume by the heart rate. Cardiac output (volume/minute) = stroke volume (volume/beat) x heart rate (beats/minute).
- **cardiorespiratory conditioning:** improvements in heart and lung function and increased blood volume, brought about by aerobic training.
- **cardiorespiratory endurance:** the ability to perform large-muscle, dynamic exercise of moderate-to-high intensity for prolonged periods.
- **conditioning:** the physical effect of training; improved flexibility, strength, and endurance.
- **cool-down:** 5 to 10 minutes of light activity, such as walking or stretching, following a vigorous workout to return the body's core gradually to near-normal temperature.
- **CP, creatine phosphate:** a high-energy compound in muscle cells that acts as a reservoir of energy that can maintain a steady supply of ATP. CP provides the energy for short bursts of activity; also called *phosphocreatine.*
- **duration:** length of time (for example, the time spent in each activity session).
- **exercise:** planned, structured, and repetitive bodily movement that promotes or maintains physical fitness.
- **fitness:** the characteristics that enable the body to perform physical activity; more broadly, the ability to meet routine physical demands with enough reserve energy to rise to a physical challenge; or the body's ability to withstand stress of all kinds.
- **flexibility:** the capacity of the joints to move through a full range of motion; the ability to bend and recover without injury.
- **frequency:** the number of occurrences per unit of time (for example, the number of activity sessions per week).
- **glucose polymers:** compounds that supply glucose, not as single molecules, but linked in chains somewhat like starch. The objective is to attract less water from the body into the digestive tract (osmotic attraction depends on the number, not the size, of particles).
- **heat stroke:** a dangerous accumulation of body heat with accompanying loss of body fluid.
- **hourly sweat rate:** the amount of weight lost plus fluid consumed during exercise per hour.
- **hyperthermia:** an above-normal body temperature.
- **hypertrophy:** growing larger; with regard to muscles, an increase in size (and strength) in response to use.
- **hyponatremia:** a decreased concentration of sodium in the blood.
- **hypothermia:** a below-normal body temperature.
- **intensity:** the degree of exertion while exercising (for example, the amount of weight lifted or the speed of running).
- **lactic acid:** the product of anaerobic glycolysis.
- **mitochondria:** the structures within a cell responsible for producing ATP.
- **moderate exercise:** activity equivalent to the rate of exertion reached when walking at a speed of 4 miles per hour (15 minutes to walk one mile).
- **muscle endurance:** the ability of a muscle to contract repeatedly without becoming exhausted.
- **muscle strength:** the ability of muscles to work against resistance.
- **physical activity:** bodily movement produced by muscle contractions that substantially increase energy expenditure.
- **progressive overload principle:** the training principle that a body system, in order to improve, must be worked at frequencies, durations, or intensities that gradually increase physical demands.
- **sedentary:** physically inactive (literally, "sitting down a lot").

- **sports anemia:** a transient condition of low hemoglobin in the blood, associated with the early stages of sports training or other strenuous activity.
- **stroke volume:** the amount of oxygenated blood the heart ejects toward the tissues at each beat.
- **training:** practicing an activity regularly, which leads to conditioning. (Training is what you do; conditioning is what you get.)
- **VO₂max:** the maximum rate of oxygen consumption by an individual at sea level.
- **warm-up:** 5 to 10 minutes of light activity, such as easy jogging or cycling, prior to a workout to prepare the body for more vigorous activity.
- **weight training**: the use of free weights or weight machines to provide resistance for developing muscle strength and endurance (also called *resistance training*). A person's own body weight may also be used to provide resistance as when a person does push-ups, pull-ups, or abdominal crunches.

Sample Test Questions

Select the best answer for each question.

1. The ability to meet routine physical demands with enough reserve energy to rise to a physical challenge is:
   a. physical activity.
   b. exercise.
   c. cardiorespiratory endurance.
   d. fitness.

2. The distinction between physical activity and exercise is:
   a. only exercise involves bodily movements and enhanced energy expenditure.
   b. physical activity benefits health more than exercise.
   c. exercise is considered more planned and structured than physical activity.
   d. exercise promotes fitness more than physical activity.

3. Lack of exercise can lead to development of:
   a. cardiovascular disease.
   b. obesity.
   c. intestinal disorders.
   d. accelerated bone losses.
   e. all of the above

4. Benefits of fitness include:
   a. restful sleep.
   b. optimal bone density.
   c. resistance to infectious diseases such as colds.
   d. strong lung function.
   e. all of the above

5. The components of fitness include:
   a. flexibility.
   b. strength.
   c. muscle endurance.
   d. cardiorespiratory endurance.
   e. all of the above

6. Muscle response to disuse or undernutrition is called:
   a. hypertrophy.
   b. atrophy.
   c. slow-twitch fibers.
   d. fast-twitch fibers.

7. Which of the following statements is true about the progressive overload principle?
   a. It refers to practicing an activity regularly.
   b. It is the ability of the muscles to work against resistance.
   c. It refers to the idea that in order to improve, training must gradually increase physical demands.
   d. It focuses on the cardiorespiratory component of fitness.

8. If a person increases the length of time performing an activity, the progressive overload principle is being applied in what way?
    a. duration
    b. frequency
    c. intensity
    d. training

9. A person who walks at a speed of 4 miles per hour is engaging in:
    a. minimal exercise.
    b. moderate exercise.
    c. vigorous training.
    d. sedentary activity.

10. The maximum rate of oxygen consumption is:
    a. aerobic.
    b. anaerobic.
    c. VO_2 max.
    d. cardiac output.

11. Lactic acid is a product of _____ metabolism.
    a. aerobic
    b. anaerobic
    c. mineral
    d. a and b
    e. a and c

12. Which of these is a high-energy compound in muscle cells that provides energy for short bursts of activity?
    a. creatine phosphate
    b. carbohydrate
    c. glycogen
    d. glycogen loading

13. Factor(s) that influence glycogen use during physical activity include:
    a. amount of carbohydrate in the diet.
    b. intensity and duration of the activity.
    c. degree of training to perform the activity.
    d. all of the above
    e. a and b

14. During low- or moderate-intensity activity, muscle cells use predominantly _____ as fuel.
    a. glycogen
    b. fat
    c. protein
    d. water
    e. none of the above

15. Which of the following statements is true about fat use for fuel?
    a. Fat can be used for fuel by muscles if the work is aerobic.
    b. Fat is preferred for fuel by athletes.
    c. Endurance athletes are encouraged to consume 40% of their energy from fat.
    d. Athletes on 30% fat diets experience fatigue.

16. Which of the following statements is true about fat use for fuel?
    a. Sustained, moderate activity uses body fat stores as its major fuel.
    b. As intensity of an activity increases, body fat contributes more fuel.
    c. People who are untrained will draw on fat more for fuel than those who are trained.
    d. People can burn fat from specific body parts.

17. Regarding protein and physical activity,
    a. the recommended protein intake for athletes is much higher than for non-athletes.
    b. protein is the primary fuel for physical activity.
    c. synthesis of body proteins is accelerated during activity.
    d. athletes retain more protein in their muscles than nonathletes.

18. Athletes can safely add muscle tissue by:
    a. tripling their protein intake.
    b. taking hormones duplicating those of puberty.
    c. putting a demand on muscles, making them work harder.
    d. depending on protein for muscle fuel and cutting down on carbohydrates.

168

19. Which factor(s) modify/ies the body's use of protein?
    a. amount of carbohydrate in the diet   d. a and b
    b. exercise intensity and duration   e. a, b, and c
    c. the degree of training

20. Which statement is true about vitamin/mineral supplements for athletes?
    a. All athletes should take them to improve performance.
    b. They should take supplements immediately prior to competition to enhance performance.
    c. A single daily multi-vitamin-mineral supplement that provides at least 10x the DRI is desirable.
    d. Consuming a well-balanced diet is preferable to taking supplements.

21. What element is important in the transport of oxygen in blood and muscle tissue and in energy transformation reactions?
    a. vitamin C   d. calcium
    b. zinc   e. thiamin
    c. iron

22. An above-normal body temperature is
    a. hyperthermia.   c. hypothermia.
    b. heat stroke.   d. sports anemia

23. The first symptom of dehydration is:
    a. collapse.   c. fatigue.
    b. hunger.   d. chills.

24. Alcohol is a diuretic and therefore it induces:
    a. energy release.   c. improved performance.
    b. fluid losses.   d. iron retention.

25. The best choice for an athlete who needs to rehydrate is:
    a. concentrated juice.   d. sweat replacers..
    b. cool water.   e. a or b
    c. salt tablets.

26. What type of athlete is prone to iron deficiency?
    a. male weight lifter   c. female endurance athlete
    b. male swimmer   d. female sprinter

27. A transient condition of low hemoglobin in the blood associated with early stages of strenuous athletic training is:
    a. sports anemia.   c. stress menstruation.
    b. menopause.   d. amenorrhea.

28. A healthy diet for athletes consists of:
    1. nutrient-dense foods.
    2. vitamin and mineral supplements.
    3. ample fluids.
    4. salt tablets.
    5. adequate food to meet energy requirements.
    6. adequate amounts of vitamins and minerals.
    7. protein powders.

    a. 1, 2, 4, 7   d. 2, 3, 4, 7
    b. 1, 3, 4, 5   e. 2, 4, 5, 6
    c. 1, 3, 5, 6

29. Alcohol hinders athletic activity by:
    a. acting as a diuretic.
    b. stimulating the central nervous system.
    c. altering perceptions.

    d. a and b
    e. a and c

30. The recommended pregame meal includes:
    a. a vitamin/mineral supplement.
    b. plenty of fluids.
    c. light, easily digestible foods.

    d. b and c
    e. a, b, and c

Short Answer Questions

1. The components of fitness are:

    a.

    b.

    c.

    d.

2. Physical activity is classified as:

    a.

    b.

3. Cardiovascular conditioning is characterized by:

    a.

    b.

    c.

    d.

    e.

    f.

Crossword Puzzle

Complete this crossword puzzle by Mary A. Wyandt, Ph.D., CHES.

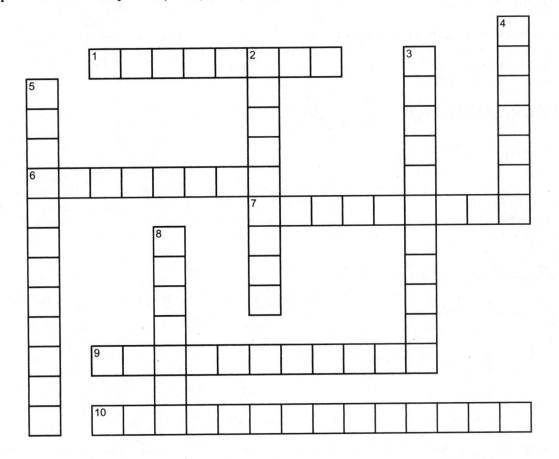

Across:	Down:
1. practicing an activity regularly, which leads to conditioning	2. the degree of exertion while exercising
6. length of time; for example, the time spent in each exercise session	3. of muscles, growing larger; an increase in size in response to use
7. physically inactive (literally, "sitting down a lot")	4. of muscles, becoming smaller; a decrease in size because of disuse, under-nutrition, or wasting disease
9. the capacity of the joints to move through a full range of motion; the ability to bend and recover without injury	5. the physical effect of training; improved flexibility, strength, and endurance
10. the ability of muscles to work against resistance	8. the characteristics that enable the body to perform physical activity; more broadly, the ability to meet routine physical demands with enough reserve energy to rise to a physical challenge; or the body's ability to withstand stress of all kinds

❀ Chapter 14 Answer Key ❀

Summing Up

1. exercise
2. fuel
3. obesity
4. insomnia
5. bone
6. strength
7. cardiovascular
8. 20
9. aerobic
10. creatine phosphate
11. glycogen
12. carbohydrate
13. duration
14. oxygen
15. supplements
16. impair
17. training
18. Glycogen
19. fluid
20. energy

Chapter Study Questions

1. Fitness: the body's ability to meet physical demands. Benefits: more restful sleep, improved nutritional health, reduced fatness and increased lean body tissue, greater bone density, improved resistance to infectious diseases, improved circulation and lung function, reduced risk of some cancers, reduced risk of diabetes, reduced incidence and severity of anxiety and depression, improved self-image and self-confidence, long life, and improved quality of life.

2. The training principle that a body system, in order to improve, must be worked at frequencies, durations, or intensities that gradually increase physical demands.

3. The ability to perform large-muscle dynamic exercise of moderate-to-high intensity for prolonged periods. Benefits include support of ongoing action of the heart and lungs.

4. Aerobic: requiring oxygen (swimming, cross-country skiing, rowing, fast walking, jogging, fast bicycling, soccer, hockey, basketball, water polo, lacrosse and rugby). Anaerobic: not requiring oxygen (jump of basketball player, weight lifting, sprinting).

5. The mixture of fuels the muscles use during physical activity depends on diet, the intensity and duration of the activity, and training. During intense activity, the fuel mix is mostly glucose (more anaerobic respiration), whereas during less intense, moderate activity, fat makes a greater contribution (more aerobic respiration). With endurance training, muscle cells adapt to store more glycogen and to rely less on glucose and more on fat for energy.

6. The body's use of glucose during physical activity depends partially on how much glycogen is in storage, and this depends partly on the amount of carbohydrate eaten. Intensity of exercise influences the body's use of glucose—high-intensity activities require more glycogen. Degree of training to perform the activity is a factor because the level of oxygen in the muscle influences the body's use of glucose. During oxygen debt, glucose is metabolized rapidly and pyruvate molecules accumulate in the muscle tissue. Duration of activity affects use of glucose during exercise—within the first 20 minutes of exercise, the body primarily uses glucose.

7. Duration of activity—when a moderate activity progresses past 20 minutes, the body uses more of its stored body fat for energy; intensity of activity—as the intensity increases, fat makes less and less of a contribution to the mixture of fuel used; degree of training—a well-trained body develops the adaptations that permit the body to draw heavily on fat for fuel.

8. Diet—people who consume diets rich in carbohydrate use less protein; intensity and duration of activity—activities more intense and long in duration will use more protein for fuel; degree of training—athletes use more protein as fuel than the unconditioned person.

9. Iron-deficiency anemia is a condition that dramatically impairs physical performance. The athlete with this condition cannot use fat for fuel or perform aerobic activities, and therefore will tire easily. Sports anemia is a transient condition of low hemoglobin in the blood, associated with the early stage of sports training or other strenuous activity. It is an adaptive, temporary response to endurance training. Iron-deficiency anemia requires iron supplementation, while sports anemia does not respond to such supplementation.

10. If the body loses too much water, its chemistry becomes compromised and dehydration may result, which can turn into heat stroke in hot, humid weather. Recommendations are to drink enough water or diluted juice before and during training and competition, rest in the shade when tired, and wear lightweight clothing.

11. A nutrient-dense diet composed mostly of unprocessed foods that meets nutrient, fluid, and energy requirements.

Sample Test Questions

1. d (p. 477)	9. b (p. 481)	17. d (p. 490)	25. b (p. 494)
2. c (p. 477)	10. c (p. 482)	18. c (p. 490)	26. c (p. 492)
3. e (p. 478)	11. b (p. 486)	19. e (p. 490-491)	27. a (p. 493)
4. e (p. 478)	12. a (p. 485)	20. d (p. 492)	28. c (p. 496-497)
5. e (p. 480)	13. d (p. 486-488)	21. c (p. 492)	29. e (p. 496)
6. b (p. 481)	14. b (p. 489)	22. a (p. 493)	30. d (p. 498)
7. c (p. 480)	15. a (p. 489)	23. c (p. 493)	
8. a (p. 480)	16. a (p. 489)	24. b (p. 496)	

Short Answer Questions

1. flexibility, strength, muscle endurance, cardiorespiratory endurance
2. aerobic, anaerobic
3. increased blood volume and oxygen delivery; increased heart strength and stroke volume; slowed resting pulse; increased breathing efficiency; improved circulation; reduced blood pressure

Crossword Puzzle

1. training	4. atrophy	7. sedentary	10. muscle strength
2. intensity	5. conditioning	8. fitness	
3. hypertrophy	6. duration	9. flexibility	

Chapter 15 – Life Cycle Nutrition: Pregnancy and Lactation

Chapter Outline

I. Nutrition Prior to Pregnancy
II. Fetal Growth and Development during Pregnancy
   A. Placental Development
   B. Fetal Growth and Development
      1. The Zygote
      2. The Embryo
      3. The Fetus
   C. Critical Periods
      1. Neural Tube Defects
      2. Folate Supplementation
      3. Chronic Diseases
      4. Fetal Programming
III. Maternal Weight
   A. Weight Prior to Conception
      1. Underweight
      2. Overweight and Obesity
   B. Weight Gain during Pregnancy
      1. Recommended Weight Gains
      2. Weight-Gain Patterns
      3. Components of Weight Gain
      4. Weight Loss after Pregnancy
   C. Exercise during Pregnancy
IV. Nutrition during Pregnancy
   A. Energy and Nutrient Needs during Pregnancy
      1. Energy
      2. Protein
      3. Essential Fatty Acids
      4. Nutrients for Blood Production and Cell Growth
      5. Nutrients for Bone Development
      6. Other Nutrients
      7. Nutrient Supplements
   B. Vegetarian Diets during Pregnancy and Lactation
   C. Common Nutrition-Related Concerns of Pregnancy
      1. Nausea
      2. Constipation and Hemorrhoids
      3. Heartburn
      4. Food Cravings and Aversions
      5. Nonfood Cravings
V. High-Risk Pregnancies
   A. The Infant's Birthweight
   B. Malnutrition and Pregnancy
      1. Malnutrition and Fertility
      2. Malnutrition and Early Pregnancy
      3. Malnutrition and Fetal Development
   C. Food Assistance Programs
   D. Maternal Health
      1. Preexisting Diabetes
      2. Gestational Diabetes
      3. Preexisting Hypertension
      4. Transient Hypertension of Pregnancy
      5. Preeclampsia and Eclampsia
   E. The Mother's Age
      1. Pregnancy in Adolescents
      2. Pregnancy in Older Women
   F. Practices Incompatible with Pregnancy
      1. Alcohol
      2. Medicinal Drugs
      3. Herbal Supplements
      4. Illicit Drugs
      5. Smoking and Chewing Tobacco
      6. Environmental Contaminants
      7. Foodborne Illness
      8. Vitamin-Mineral Megadoses
      9. Caffeine
      10. Weight-Loss Dieting
      11. Sugar Substitutes
VI. Nutrition during Lactation
   A. Lactation: A Physiological Process
   B. Breastfeeding: A Learned Behavior
   C. Maternal Energy and Nutrient Needs during Lactation
      1. Energy Intake and Exercise
      2. Energy Nutrients
      3. Vitamins and Minerals
      4. Water
      5. Nutrient Supplements
      6. Food Assistance Programs
      7. Particular Foods
   D. Maternal Health
      1. HIV Infection and AIDS
      2. Diabetes
      3. Postpartum Amenorrhea
      4. Breast Health
   E. Practices Incompatible with Lactation
      1. Alcohol
      2. Medicinal Drugs
      3. Illicit Drugs
      4. Smoking
      5. Environmental Contaminants
      6. Caffeine

VII. Fetal Alcohol Syndrome
    A. Drinking during Pregnancy

B. How Much Is Too Much?
C. When Is the Damage Done?

Summing Up

(1)_____ is a major factor influencing the nutritional needs of developing infants and children. The growth rate is fastest during (2)_____ life and the first year. During pregnancy, changes in both mothers' and infants' bodies necessitate increased intakes of the (3)_____ nutrients. A pregnant woman should gain about (4)_____ pounds from foods of high nutrient density. Malnutrition during pregnancy affects the developing fetus; (5)_____ babies often fail to thrive. (6)_____, smoking, drugs, dieting, and unbalanced nutrient intakes of all kinds should be avoided for the duration of pregnancy. (7)_____ and ample food energy are especially important. Fluid intake should be liberal and (8)_____ normally should not be restricted.

The breastfeeding mother needs additional (9)_____ from foods of high nutrient density and a generous fluid intake. The rapidly growing newborn infant requires milk, preferably (10)_____ milk, which provides the needed nutrients in quantities suitable to support the infant's growth. Advantages of breast milk over formula, especially in (11)_____ countries, are that it protects the infant against (12)_____ and that it is sanitary, economical, and premixed to the correct proportions. To avoid (13)_____, all susceptible infants should be breastfed at first, even in developed countries.

Chapter Study Questions

1.   Describe the placenta and its function.

2.   Describe the normal events of fetal development. How does malnutrition impair fetal development?

3.   Define the term critical period. How do adverse influences during critical periods affect later health?

4.   Explain why women of childbearing age need folate in their diets. How much is recommended, and how can women ensure that these needs are met?

5. What is the recommended pattern of weight gain during pregnancy for a woman at a healthy weight? For an underweight woman? For an overweight woman? For an obese woman?

6. What does a pregnant woman need to know about exercise?

7. Which nutrients are needed in the greatest amounts during pregnancy? Why are they so important? Describe wise food choices for the pregnant woman.

8. Define low-risk and high-risk pregnancies. What is the significance of infant birthweight in terms of the child's future health?

9. Describe some of the special problems of the pregnant adolescent. Which nutrients are needed in increased amounts?

10. What practices should be avoided during pregnancy? Why?

11. How do nutrient needs during lactation differ from nutrient needs during pregnancy?

Chapter Glossary

- **amniotic sac:** the "bag of waters" in the uterus, in which the fetus floats.
- **anencephaly:** an uncommon and always fatal type of neural tube defect, characterized by the absence of a brain.
- **appropriate for gestational age (AGA):** infants born at a size and weight appropriate for their length of gestation.

- **cesarean section:** a surgically assisted birth involving removal of the fetus by an incision into the uterus, usually by way of the abdominal wall.
- **conception:** the union of the male sperm and the female ovum; fertilization.
- **critical periods:** finite periods during development in which certain events may occur that will have irreversible effects on later developmental stages; usually a period of rapid cell division.
- **Down syndrome:** a genetic abnormality that causes mental retardation, short stature, and flattened facial features.
- **eclampsia:** a severe stage of preeclampsia characterized by convulsions.
- **embryo:** the developing infant from two to eight weeks after conception.
- **fertility:** the capacity of a woman to produce a normal ovum periodically and of a man to produce normal sperm; the ability to reproduce.
- **fetal programming**: the influences of substances during fetal growth on the development of diseases in later life.
- **fetus:** the developing infant from eight weeks after conception until term.
- **food aversions:** strong desires to avoid particular foods.
- **food cravings:** strong desires to eat particular foods.
- **gestation:** the period from conception to birth. For human beings, the average length of a healthy gestation is 40 weeks. Pregnancy is often divided into thirds, called *trimesters*.
- **gestational diabetes:** abnormal glucose tolerance during pregnancy.
- **high-risk pregnancy:** a pregnancy characterized by indicators that make it likely the birth will be surrounded by problems such as premature delivery, difficult birth, retarded growth, birth defects, and early infant death.
- **implantation:** the stage of development in which the zygote embeds itself in the wall of the uterus and begins to develop; occurs during the first two weeks after conception.
- **lactation:** production and secretion of breast milk for the purpose of nourishing an infant.
- **let-down reflex:** the reflex that forces milk to the front of the breast when the infant begins to nurse.
- **listeriosis:** an infection caused by eating food contaminated with the bacterium *Listeria monocytogenes*, which can be killed by pasteurization and cooking, but can survive at refrigerated temperatures; certain ready-to-eat foods, such as hot dogs and deli meats, may become contaminated after cooking or processing, but before packaging.
- **low birthweight (LBW):** a birthweight of 5 ½ pounds (2500 grams) or less; indicates probable poor health in the newborn and poor nutrition status in the mother during pregnancy, before pregnancy, or both. Normal birthweight for a full-term baby is 6 ½ to 8 ¾ pounds (about 3000 to 4000 g).
- **low-risk pregnancy:** a pregnancy characterized by indicators that make a normal outcome likely.
- **macrosomia:** describes high-birthweight infants (roughly 9 lb., or 4000 g, or more) resulting from prepregnancy obesity, excessive weight gain during pregnancy, or uncontrolled diabetes.
- **mammary glands**: glands of the female breast that secrete milk.
- **neural tube defect:** a malformation of the brain, spinal cord, or both during embryonic development.
- **ovum:** the female reproductive cell, capable of developing into a new organism upon fertilization; commonly referred to as an egg.
- **oxytocin:** a hormone that stimulates the mammary glands to eject milk during lactation and the uterus to contract during childbirth.
- **placenta:** the organ that develops inside the uterus early in pregnancy, through which the fetus receives nutrients and oxygen and returns carbon dioxide and other waste products to be excreted.
- **postpartum amenorrhea:** the normal temporary absence of menstrual periods immediately following childbirth.
- **post term infant:** an infant born after the 42nd week of pregnancy.
- **preeclampsia:** a condition characterized by hypertension, fluid retention, and protein in the urine; formerly known as *pregnancy-induced hypertension*.
- **preterm infant:** an infant born prior to the 38th week of pregnancy; also called a *premature infant*. A *term* infant is born between the 38th and 42nd week of pregnancy.
- **prolactin:** a hormone secreted from the anterior pituitary gland that acts on the mammary glands to promote (*pro*) the production of milk (*lacto*). The release of prolactin is mediated by *prolactin-inhibiting hormone (PIH)*.

- **small for gestational age (SGA):** an infant who has suffered growth failure in the uterus, and is thus smaller than expected based on the length of gestation.
- **sperm:** the male reproductive cell, capable of fertilizing an ovum.
- **spina bifida:** one of the most common types of neural tube defects; characterized by the incomplete closure of the spinal cord and its bony encasement.
- **sudden infant death syndrome (SIDS):** the unexpected and unexplained death of an apparently well infant; the most common cause of death of infants between the second week and the end of the first year of life; also called *crib death*.
- **teratogenic:** describes a factor that causes abnormal fetal development and birth defects.
- **toxemia:** the hypertensive diseases of pregnancy.
- **transient hypertension of pregnancy:** high blood pressure that develops in the second half of pregnancy and resolves after childbirth, usually without affecting the outcome of the pregnancy.
- **umbilical cord:** the ropelike structure through which the fetus's veins and arteries reach the placenta; the route of nourishment and oxygen to the fetus and the route of waste disposal from the fetus.
- **umbilicus:** the scar in the middle of the abdomen that marks the former attachment of the umbilical cord, commonly known as the "belly button."
- **uterus:** the muscular organ within which the infant develops before birth.
- **zygote:** the product of the union of ovum and sperm; so-called for the first two weeks after fertilization.

Sample Test Questions

Select the best answer for each question.

1. The capacity of men and women to reproduce is:
    a. implantation.
    b. gestation.
    c. fertility.
    d. ferrum.

2. The union of the male sperm and female ovum is called:
    a. a zygote.
    b. conception.
    c. festering.
    d. a placenta.

3. The product of the union of the male sperm and female ovum is called:
    a. a zygote.
    b. conception.
    c. fertilization.
    d. a fetus.

4. The muscular organ within which the infant develops before birth is the:
    a. placenta.
    b. amniotic sac.
    c. vagina.
    d. uterus.

5. The structure through which the fetus's veins and arteries reach the placenta is the:
    a. umbilical cord.
    b. ovum.
    c. amniotic sac.
    d. fallopian tubes.

6. In preparation for a healthy pregnancy, women can establish the following habits:
    a. maintain a healthy body weight and be physically active.
    b. consume an adequate and healthy diet, and receive regular medical care.
    c. select high levels of vitamin/mineral supplements and obtain colon cleansings.
    d. a and b
    e. a and c

7. Nutrients and oxygen travel to the developing fetus via:
    a. the placenta.
    b. the amniotic sac.
    c. its lungs.
    d. its intestines.

8. The growth period when a lack of nutrients is most likely to produce permanent change is the:
   a. period of rapid increase in number of cells.
   b. period of rapid increase in size of cells.
   c. teen growth spurt.
   d. growth spurt from 1 to 20 years of age.

9. A fetus is the:
   a. developing infant during its second through eighth week after conception.
   b. developing infant from the eighth week after conception until birth.
   c. muscular organ within which the infant develops before birth.
   d. infant from birth until its first birthday.
   e. organ from which the infant receives nourishment.

10. The critical period for neural tube development is from:
    a. 17 to 30 days gestation.   c. 41 to 60 days gestation.
    b. 31 to 40 days gestation.   d. 61 to 70 days gestation.

11. Neural tube defects result in the following condition(s):
    a. spina bifida.   d. a and c
    b. anencephaly.   e. a and b
    c. amorphous.

12. A pregnant woman (compared to a non-pregnant woman) is advised to consume almost twice as much/many:
    a. kcalories.   d. carbohydrate.
    b. iron.   e. folate.
    c. fat.

13. Which of the following statements is true regarding pregnancy and chronic diseases?
    a. Adverse influences at critical times are not related to chronic diseases later in life.
    b. Poor maternal diet may influence blood pressure but not immune functions.
    c. Malnutrition is not an influence on type 2 diabetes.
    d. Low birth weight and premature birth correlate with insulin resistance.

14. The influence of substances during fetal growth on the development of diseases later in life is:
    a. gestational diabetes.   c. high-risk pregnancy.
    b. fetal programming.   d. Down syndrome.

15. An infant born prior to the 38th week of pregnancy is:
    a. term.   c. preterm.
    b. post term.   d. low risk.

16. A surgically assisted birth involving an incision into the uterus is called:
    a. vaginal delivery.   c. eclampsia.
    b. cesarean section.   d. episiotomy.

17. Strong desires to avoid certain foods are:
    a. food cravings.   c. food allergies.
    b. food intolerances.   d. food aversions.

18. High-birthweight infants are called:
    a. post term.   c. macroscelian.
    b. macrosomic.   d. macrogametes.

19. For women who begin pregnancy at a healthy weight, physicians often recommend:
    a. weight gain of about 25 to 35 pounds.
    b. no mineral or vitamin supplements and weight gain of at least 35 pounds.
    c. vitamin C supplements and weight gain of about 20 pounds.
    d. multi-vitamin and -mineral supplements and weight gain of no more than 20 pounds.

20. Because of the dangers of obesity, an overweight pregnant woman should try to gain:
    a. 25 to 28 pounds during pregnancy.   c. 15 to 25 pounds during pregnancy.
    b. 28 to 40 pounds during pregnancy.   d. no weight during pregnancy.

21. A severe stage of hypertension, fluid retention, protein in the urine, and convulsions is:
    a. gestational diabetes.   c. eclampsia.
    b. preeclampsia.   d. SIDS.

22. Which of the following statements is true regarding nutrition during pregnancy?
    a. Damage done to an infant by a woman's alcohol abuse during pregnancy can be corrected after birth by proper nutrition.
    b. The AI for vitamin D increases during pregnancy.
    c. The iron RDA during pregnancy is 27 mg/day.
    d. Calcium absorption and retention during pregnancy decrease.

23. Vegan diets during pregnancy:
    a. always result in low-birthweight babies that have statistically greater chances of contracting diseases and of dying early in life.
    b. do not require any supplementation.
    c. require iron supplementation.
    d. do not require calcium supplementation.

24. The stage of development in which the fertilized egg embeds itself in the wall of the uterus is called:
    a. the ovum.   d. the zygote.
    b. the critical period.   e. implantation.
    c. the placenta.

25. A pregnant woman who is constipated asks a health care provider for advice. Assuming no disease is present, what should the health care provider probably tell her to try first?
    a. increase fluid intake   c. eat fewer bulky foods
    b. use laxatives   d. reduce fluid intake

26. During the second trimester, a pregnant woman needs _____ kcalories above the allowance for a nonpregnant woman.
    a. 340   c. 530
    b. 450   d. 620

27. The protein RDA for pregnancy is _____ grams per day higher than for nonpregnant women.
    a. 2   c. 10
    b. 6   d. 25

28. The increased need for _____ in pregnancy is difficult to meet by diet or by existing stores; therefore supplements are recommended.
    a. vitamin C   c. iron
    b. sodium   d. vitamin D

29. The most common outcome of a high-risk pregnancy is:
    a. diabetes.   c. weight appropriate for gestational age.
    b. heart disease.   d. low birthweight.

30. Pregnant teenagers have high rates of
    a. stillbirths.
    b. preterm births.
    c. low-birthweight infants.

    d. all of the above
    e. b and c

Short Answer Questions

1. List conditions that raise risk in pregnancy:

    a.

    b.

    c.

    d.

    e.

    f.

    g.

Problem Solving

1. What is the recommended energy intake during the 2nd and 3rd trimesters of pregnancy for a woman at a healthy weight whose non-pregnancy estimated energy expenditure is 1800 kcalories?

2. Ideally, how much weight would you expect a woman to have gained by the 32nd week of pregnancy?

Crossword Puzzle

Complete this crossword puzzle by Mary A. Wyandt, Ph.D., CHES.

Across:		Down:	
1.	the period from conception to birth; for humans it lasts from 38 to 42 weeks	2.	the female reproductive cell
4.	the male reproductive cell	3.	the developing infant from eight weeks after conception until term
7.	the muscular organ within which the infant develops before birth; the womb	5.	the organ that develops inside the uterus early in pregnancy, in which maternal and fetal blood circulate in close proximity so that materials can be exchanged between them
9.	the capacity of a woman to produce a normal ovum periodically and of a man to produce normal sperm; the ability to reproduce	6.	the product of the union of ovum and sperm; so-called for the first two weeks after fertilization
10.	the union of the male sperm and the female ovum; fertilization	8.	the developing infant from two to eight weeks after conception

✤ Chapter 15 Answer Key ✤

Summing Up

1. Growth
2. prenatal
3. growth
4. 25 to 35
5. low-birthweight
6. Alcohol
7. Protein
8. salt
9. kcalories
10. breast
11. developing
12. disease
13. allergy

Chapter Study Questions

1. The placenta is the organ that develops inside the uterus early in pregnancy, in which maternal and fetal blood circulate in close proximity so that materials can be exchanged between them. The baby receives nutrients and oxygen across the placenta, and the mother's blood picks up carbon dioxide and other waste products to be excreted.

2. During the 7 months of fetal development, each organ grows to maturity according to its own characteristic schedule, with greater intensity at some times than at others. Intense development and rapid cell division take place only during certain times. During times of critical periods each organ and tissue is most vulnerable to an insult such as a nutrient deficiency. If nutrients are not available during these times, significant damage can take place with regard to organ development.

3. Critical periods are finite periods during development in which certain events may occur that will have irreversible effects on later developmental stages. In the case of the developing fetus, a critical period is a period of rapid cell division. Early malnutrition during these times affects the heart and lungs in later life.

4. Folate can prevent neural tube defects. 0.4 mg/day is recommended; women should eat plenty of fruits and vegetables or take supplements.

5. Underweight: 28 to 40 lb.; normal-weight: 25 to 35 lb.; overweight: 15 to 25 lb.; obese: 15 lb. minimum.

6. A woman who is active prior to pregnancy and is experiencing a normal pregnancy can continue to exercise throughout her pregnancy, but will want to avoid sports in which she might fall or be hit by other people or objects such as racquetball. Women are advised to avoid overly strenuous activities and to abstain from exercising in hot weather, and to stay out of saunas, steam rooms, and hot whirlpools.

7. Energy nutrients are essential (a women needs extra food energy to support the growth and metabolic activities of the placenta, fetus, and her own body tissues), extra protein is needed for growth of new tissues, additional B vitamins are needed in proportion to increased food energy intake (they function as coenzymes in energy reactions), folate and B_{12} are needed for blood cells and growth, vitamin D and minerals are necessary for bone development, extra iron is needed to provide for fetal and placental needs and usually to boost pre-pregnancy low blood levels, and zinc is needed for DNA and RNA synthesis (thus for protein synthesis and cell development). Wise food choices include a balance similar to that of the USDA Food Guide with additional servings for each of the five food groups to meet increased nutrient needs (based on amounts recommended for the higher kcal intakes during the second and third trimesters).

8. Low-risk pregnancy: a pregnancy characterized by indicators that make a normal outcome likely; high-risk pregnancy: a pregnancy characterized by indicators that make it likely the birth will be surrounded by problems such as premature delivery, difficult birth, retarded growth, birth defects, and early infant death. Birthweight is a potential predictor of an infant's future health and survival.

9. A teenaged female is growing and has to meet her own nutrient needs, in addition to trying to meet the nutrient needs of a developing baby. The demands of pregnancy compete with those of her own growth, placing her and the infant at high risk for complications. Protein and vitamin K are needed in increased amounts.

10. Drinking alcohol, taking medicinal drugs, using illicit drugs, smoking or using smokeless tobacco, ingesting foods and beverages contaminated with lead or mercury, taking vitamin-mineral megadoses, drinking caffeine, weight-loss dieting. Substances can pass to the fetus and cause irreversible harm or death.

11. In general, nutrient needs are as high or higher during lactation as during pregnancy (except protein and folate).

Sample Test Questions

1. c (p. 509)	9. b (p. 512)	17. d (p. 525)	25. a (p. 524)
2. b (p. 510)	10. a (p. 512)	18. b (p. 516)	26. a (p. 520)
3. a (p. 510)	11. e (p. 513)	19. a (p. 516)	27. d (p. 520)
4. d (p. 510)	12. e (p. 514)	20. c (p. 516-517)	28. c (p. 522)
5. a (p. 510)	13. d (p. 515)	21. c (p. 528)	29. d (p. 525)
6. d (p. 509)	14. b (p. 515)	22. c (p. 522)	30. d (p. 529)
7. a (p. 510)	15. c (p. 516)	23. c (p. 522, 524)	
8. a (p. 512)	16. b (p. 516)	24. e (p. 510)	

Short Answer Questions

1. alcohol consumption; use of medicinal drugs, herbal supplements, illicit drugs, tobacco; exposure to environmental contaminants; experiencing foodborne illness; consuming vitamin/mineral megadoses; consuming caffeine; weight loss dieting

Problem Solving

1. 1800 + 340 kcalories per day during 2nd trimester = 2140 kcalories
   1800 + 450 kcalories per day during 3rd trimester = 2250 kcalories

2. 3.5 lb./1st 3 mo (13 wk) + 19 wk x 1 lb./week = 3.5 + 19 = 22.5 lb.

Crossword Puzzle

1. gestation	4. sperm	7. uterus	10. conception
2. ovum	5. placenta	8. embryo	
3. fetus	6. zygote	9. fertility	

⑤ Chapter 16 – Life Cycle Nutrition: ⑥ Infancy, Childhood, and Adolescence

Chapter Outline

I. Nutrition during Infancy
  A. Energy and Nutrient Needs
    1. Energy Intake and Activity
    2. Energy Nutrients
    3. Vitamins and Minerals
    4. Water
  B. Breast Milk
    1. Frequency and Duration of Breastfeeding
    2. Energy Nutrients
    3. Vitamins
    4. Minerals
    5. Supplements
    6. Immunological Protection
    7. Allergy and Disease Protection
    8. Other Potential Benefits
    9. Breast Milk Banks
  C. Infant Formula
    1. Infant Formula Composition
    2. Risks of Formula Feeding
    3. Infant Formula Standards
    4. Special Formulas
    5. Inappropriate Formulas
    6. Nursing Bottle Tooth Decay
  D. Special Needs of Preterm Infants
  E. Introducing Cow's Milk
  F. Introducing Solid Foods
    1. When to Begin
    2. Food Allergies
    3. Choice of Infant Foods
    4. Foods to Provide Iron
    5. Foods to Provide Vitamin C
    6. Foods to Omit
    7. Vegetarian Diets during Infancy
    8. Foods at One Year
  G. Mealtimes with Toddlers
II. Nutrition during Childhood
  A. Energy and Nutrient Needs
    1. Energy Intake and Activity
    2. Carbohydrate and Fiber
    3. Fat and Fatty Acids
    4. Protein
    5. Vitamins and Minerals
    6. Supplements
    7. Planning Children's Meals
  B. Hunger and Malnutrition in Children
    1. Hunger and Behavior
    2. Iron Deficiency and Behavior

    3. Other Nutrient Deficiencies and Behavior
  C. The Malnutrition-Lead Connection
  D. Hyperactivity and "Hyper" Behavior
    1. Hyperactivity
    2. Misbehaving
  E. Food Allergy and Intolerance
    1. Detecting Food Allergy
    2. Anaphylactic Shock
    3. Food Labeling
    4. Food Intolerances
  F. Childhood Obesity
    1. Genetic and Environmental Factors
    2. Growth
    3. Physical Health
    4. Psychological Development
    5. Prevention and Treatment of Obesity
    6. Diet
    7. Physical Activity
    8. Psychological Support
    9. Behavioral Changes
  G. Mealtimes at Home
    1. Honoring Children's Preferences
    2. Learning through Participation
    3. Avoiding Power Struggles
    4. Choking Prevention
    5. Playing First
    6. Snacking
    7. Preventing Dental Caries
    8. Serving as Role Models
  H. Nutrition at School
    1. Meals at School
    2. Competing Influences at School
III. Nutrition during Adolescence
  A. Growth and Development
  B. Energy and Nutrient Needs
    1. Energy Intake and Activity
    2. Vitamins
    3. Iron
    4. Calcium
  C. Food Choices and Health Habits
    1. Snacks
    2. Beverages
    3. Eating Away from Home
    4. Peer Influence
  D. Problems Adolescents Face
    1. Marijuana
    2. Cocaine

3. Ecstasy
4. Drug Abuse, in General
5. Alcohol Abuse
6. Smoking
7. Smokeless Tobacco

IV. Childhood Obesity and the Early Development
of Chronic Diseases
A. Early Development of Type 2 Diabetes
B. Early Development of Heart Disease

1. Atherosclerosis
2. Blood Cholesterol
3. Blood Pressure
C. Physical Activity
D. Dietary Recommendations for Children
1. Moderation, Not Deprivation
2. Diet First, Drugs Later
E. Smoking

Summing Up

After the age of one, a child's growth rate (1)_____, and with it, the appetite. However, all essential nutrients continue to be needed in adequate amounts from foods with a high nutrient (2)_____.

When children go to school, their nutrition needs are partly met by school (3)_____ programs. Another influential factor in the lives of children is (4)_____, with many advertisements for sugary foods; another is (5)_____ machines, which often limit choices to foods of low quality. (6)_____ and other health professionals are concerned that the advertisement and availability of sugary foods should be controlled; there may also be a need to control some children's (7)_____ consumption from cola beverages and cocoa products.

Sound nutrition practices may prevent future health problems to some extent—among them, (8)_____, iron-deficiency anemia, (9)_____ disease, and diabetes. Screening for these conditions ensures early detection and facilitates early control.

It is desirable for children to learn to like (10)_____ foods in all the food groups. This liking seems to come naturally except, in some children, the liking for (11)_____. The person who feeds the child must be aware of the child's (12)_____ and emotional development. Children can learn positive eating (13)_____ that will continue to promote their good health after they have become adults.

The (14)_____ years mark the transition from a time when children eat what they are fed to a time when they choose for themselves what to eat. Nutrition (15)_____ becomes important as a means of encouraging healthy food habits. Teenagers' (16)_____ patterns and lifestyles predispose them to certain nutrient inadequacies, notably a lack of (17)_____, but teenagers vary so widely that generalizations are difficult. Special problems that may arise in the teen years and affect nutrition include pregnancy and alcohol, tobacco and (18)_____ use.

Chapter Study Questions

1. Describe some of the nutrient and immunological attributes of breast milk.

2. What are the appropriate uses of formula feeding? What criteria would you use in selecting an infant formula?

3. Why are solid foods not recommended for an infant during the first few months of life? When is an infant ready to start eating solid food?

4. Identify foods that are inappropriate for infants and explain why they are inappropriate.

5. What nutrition problems are most common in children? What strategies can help prevent them?

6. Describe the relationships between nutrition and behavior. How does television influence nutrition?

7. Describe a true food allergy. Which foods most often cause allergic reactions? How do food allergies influence nutrition status?

8. Describe the problems associated with childhood obesity and the strategies for prevention and treatment.

9. List strategies for introducing nutritious foods to children.

10. What impact do school meal programs have on the nutrition status of children?

11. Describe changes in nutrient needs from childhood to adolescence. Why is a teenaged girl more likely to develop an iron deficiency than is a boy?

12. How do adolescents' eating habits influence their nutrient intakes?

13. How does the use of illicit drugs influence nutrition status?

14. How do the nutrient intakes of smokers differ from those of nonsmokers? What impacts can those differences exert on health?

Chapter Glossary

- **adolescence:** the period from the beginning of puberty until maturity.
- **adverse reactions:** unusual responses to food (including intolerances and allergies).
- **alpha-lactalbumin:** a major protein in human breast milk, as opposed to *casein*, a major protein in cow's milk.
- **anaphylactic shock:** a life-threatening whole-body allergic reaction to an offending substance.
- **asymptomatic allergy:** allergy in which an individual produces antibodies without symptoms.
- **beikost:** any nonmilk foods given to an infant.
- **bifidus factors:** factors in colostrum and breast milk that favor the growth of the "friendly" bacterium *Lactobacillus bifidus* in the infant's intestinal tract, so that other, less desirable intestinal inhabitants will not flourish.
- **botulism:** an often fatal food-borne illness caused by the ingestion of foods containing a toxin produced by bacteria that grow without oxygen.
- **breast milk bank:** a service that collects, screens, processes, and distributes donated human milk.
- **colostrum:** a milklike secretion from the breast, present during the first day or so after delivery before milk appears; rich in protective factors.
- **epinephrine:** a hormone of the adrenal gland that modulates the stress response; formerly called *adrenaline*. When administered by injection, epinephrine counteracts anaphylactic shock by opening the airways and maintaining heartbeat and blood pressure.
- **food allergy:** an adverse reaction to food that involves an immune response; also called *food-hypersensitivity reaction*.
- **food intolerances:** adverse reactions to foods that do not involve the immune system.
- **gatekeepers:** with respect to nutrition, key people who control other people's access to foods and thereby exert profound impacts on their nutrition. Examples are the spouse who buys and cooks the food, the parent who feeds the children, and the caregiver in a day-care center.
- **hyperactivity:** inattentive and impulsive behavior that is more frequent and severe than is typical of others at a similar age; professionally called *attention deficit/hyperactivity disorder (ADHD)*.

- **hypoallergenic formulas:** clinically tested infant formulas that do not provoke reactions in 90% of infants or children with confirmed cow's milk allergy. Like all infant formulas, hypoallergenic formulas must demonstrate nutritional suitability to support infant growth and development. Extensively hydrolyzed and free amino acid-based formulas are examples.
- **lactadherin:** a protein in breast milk that attacks diarrhea-causing viruses.
- **lactoferrin:** a protein in breast milk that binds iron and keeps it from supporting the growth of the infant's intestinal bacteria.
- **milk anemia:** iron-deficiency anemia that develops when an excessive milk intake displaces iron-rich foods from the diet.
- **nursing bottle tooth decay:** extensive tooth decay due to prolonged contact with formula, milk, fruit juice, or other carbohydrate-rich liquid offered to an infant in a bottle.
- **puberty:** the period in life in which a person becomes physically capable of reproduction.
- **serotonin:** a neurotransmitter important in the regulation of appetite, sleep, and body temperature.
- **symptomatic allergy:** allergy in which an individual produces antibodies and has symptoms.
- **tolerance level:** the maximum amount of residue permitted in a food when a pesticide is used according to the label directions.
- **wean:** gradually replacing breast milk with infant formula or other foods appropriate to an infant's diet.

Sample Test Questions

Select the best answer for each question.

1. Which of the following statements is true?
   a. After the age of one, a child's growth rate accelerates.
   b. Lactadherin is the major protein in breast milk.
   c. Colostrum is a secretion from the breast, rich in protective factors.
   d. Bifidus factors are undesirable.

2. A protein in breast milk that binds iron and keeps it from supporting the growth of the infant's intestinal bacteria is:
   a. casein.
   b. lactoferrin.
   c. lactadherin.
   d. alpha-lactalbumin.

3. The American Academy of Pediatrics recommends that:
   a. infants exclusively receive breast milk for the first 6 months.
   b. infants receive breast milk and complementary foods for 3 to 6 months.
   c. infants receive breast milk or formula for the first 12 months.
   d. infants exclusively receive breast milk for the first 3 years.

4. Since breast milk contains relatively low amounts of iron,
   a. iron supplementation may be suggested.
   b. formula feeding is preferred.
   c. iron has a low bioavailability.
   d. feeding liver to the infant is suggested.

5. Benefits of breast feeding include:
   a. immunological protection including colostrum that contains antibodies.
   b. bifidus factors that favor the growth of friendly bacteria in the infant's digestive tract.
   c. lactoferrin and lactadherin that protect against infant diarrhea.
   d. protection against allergy development.
   e. all of the above

6. At what age can a normal infant begin eating solid foods?
   a. 3-5 weeks
   b. 26-32 weeks
   c. 4-6 months
   d. 9-12 months

7. What should be the first food introduced to the infant?
   a. yogurt
   b. egg white
   c. rice cereal
   d. finely chopped meat

8. The main purpose of introducing solid food is:
   a. to help the infant sleep through the night.
   b. to provide nutrients that are no longer supplied adequately by breast milk alone.
   c. to increase body weight.
   d. to improve mental capacity.

9. Which of the following changes in body structure usually takes place between the ages of 1 and 2 years?
   a. weight doubles
   b. weight triples
   c. length of body doubles
   d. length of long bones increases

10. Approximately how many kcalories per day does an average 3-year old need to obtain?
    a. 500
    b. 800
    c. 1300
    d. 2400

11. Which of the following provides the most important information about a child's health?
    a. growth chart
    b. blood lipid profile
    c. long-bone size and density
    d. onset of walking and talking

12. Two to six year olds need _____ servings from the milk group daily.
    a. one
    b. one-half
    c. two
    d. three

13. A large amount of concentrated sweets in a child's diet is most likely to lead to:
    a. apathy.
    b. obesity.
    c. hyperactivity.
    d. growth inhibition.

14. An adverse reaction to foods that does not involve an immune response is:
    a. a food intolerance.
    b. a food allergy.
    c. an antigen.
    d. anaphylactic.

15. Inattentive and impulsive behavior that is more frequent and severe than is typical of others of a similar age is:
    a. hypoallergenic.
    b. hyperactivity.
    c. obsessive-compulsive disorder.
    d. adversive behavior.

16. How much more total energy does a normal 10 year old need compared with a normal 1 year old?
    a. 25%
    b. 50%
    c. 100%
    d. 200%

17. Which of the following two conditions are associated with television's influence?
    a. obesity and poor dental health
    b. drug abuse and teenage pregnancy
    c. anorexia and nutrient deficiencies
    d. hyperactivity and lower body weight

18. Compared with adolescent boys in terms of height, weight, and body composition, adolescent girls:
    a. start growing later.
    b. lay down more fat.
    c. have more lean body mass.
    d. have more energy.

19. The School Lunch Program is intended to provide at least _____ of children's RDA for each of several nutrients.
    a. one fourth
    b. one third
    c. one hundred percent
    d. No requirement stipulated.

20. Which of the following statements is true regarding obesity?
    a. The number of overweight children has remained steady over the past three decades.
    b. Parental obesity is not correlated with childhood obesity.
    c. Diet and physical activity are unrelated to childhood obesity.
    d. Soft drink intake and fast food consumption play roles in childhood weight gains.

21. Key people who control other people's access to foods are called:
    a. gatekeepers.
    b. food monitors.
    c. chiefs.
    d. food directors.

22. One nutrient that often comes up short in teenagers' diets is:
    a. protein.
    b. potassium.
    c. thiamin.
    d. iron.
    e. riboflavin.

23. A negative influence in schools regarding nutrition is:
    a. the School Lunch Program.
    b. the School Breakfast Program.
    c. foods from vending machines.
    d. all of the above

24. Adolescents who drink more soft drinks and less milk
    a. have high intakes of iron.
    b. have high intakes of calcium.
    c. have high intakes of fat.
    d. have low energy intakes.
    e. are more likely to be overweight.

25. The single most effective way to teach nutrition to children is by:
    a. example.
    b. punishment.
    c. singling out only hazardous nutrition practices for attention.
    d. explaining the importance of eating new foods as a prerequisite for dessert.

26. What can parents do to help their children consume a balanced diet?
    a. Allow eating only at mealtimes and forbid all snacking.
    b. Make a variety of nutritious foods available.
    c. Insist on keeping close track of everything they eat.
    d. Parents really can't do anything to influence their diets.

27. A teenager's kcalorie intake from snacks averages about _____ percent of total daily intake.
    a. five
    b. fifteen
    c. twenty-five
    d. forty
    e. fifty

28. A nutrition problem observed in people with drug addictions is:
    a. they spend money for drugs that could be spent on food.
    b. they lose interest in food during highs.
    c. some drugs depress the appetite.
    d. all of the above

29. A problem with alcohol consumption among adolescents is:
    a. It provides nutrients with little energy.
    b. It can displace nutritious foods from the diet.
    c. It hinders nutrient absorption and metabolism.
    d. a and b
    e. b and c

30. A problem with smoking and nutrition is:
    a. smokers require less vitamin C than nonsmokers.
    b. smokers consume greater amounts of fruits and vegetables than nonsmokers.
    c. smokers have lower intakes of fiber, vitamin A, beta-carotene, folate, and vitamin C.
    d. smoking causes weight gain.

Short Answer Questions

1. List iron-rich foods that most children like.

2. List 5 physical signs of malnutrition in children.

    a.

    b.

    c.

    d.

    e.

Crossword Puzzle

Complete this crossword puzzle by Mary A. Wyandt, Ph.D., CHES.

	Across:		Down:
1.	a protein in breast milk that attacks diarrhea-causing viruses	2.	the chief protein in human breast milk
7.	adverse reactions to foods that do not involve the immune system	3.	a factor in breast milk that binds iron and keeps it from supporting the growth of the infant's intestinal bacteria
8.	a milk-like secretion from the breast, present during the first day or so after delivery before milk appears; rich in protective factors	4.	supplemental or weaning foods
		5.	to gradually replace breast mild with infant formula or other foods appropriate to an infant's diet
9.	an often fatal food-borne illness caused by ingesting foods that contain a toxin produced by bacteria that grow without oxygen	6.	the period from the beginning of puberty until maturity
10.	a metabolic bone disease common in preterm infants		

☾ Chapter 16 Answer Key ☾

Summing Up

1. declines
2. density
3. lunch
4. television
5. vending
6. Dentists
7. caffeine
8. obesity
9. cardiovascular
10. nutritious
11. vegetables
12. psychological
13. habits
14. teen
15. education
16. snacking
17. iron
18. drug

Chapter Study Questions

1. Breast milk provides natural antibodies, it is sterile, and it is nutritionally tailor-made to meet infant's needs.
2. To substitute for breast milk occasionally, or to wean to formula during the first year. Formula must meet AAP standards.
3. Allergy development, physical immaturity; the infant's nutrient needs are met by body stores and formula or breast milk until then. Infant is ready to start eating solid food when the infant's birth weight has doubled, the infant consumes 8 ounces of formula and gets hungry again in less than 4 hours, the infant can sit up, the infant consumed 32 ounces a day and wants more, and the infant is six months old.
4. Ordinary milk because it provides insufficient vitamin C and iron and excessive sodium and protein; mixed dinners and heavily sweetened desserts because they are not nutrient dense; sweets of any kind including baby food "desserts" because they convey no nutrients to support growth and may promote obesity; canned vegetables because they contain too much sodium; honey and corn syrup because of the risk of botulism; popcorn, whole grapes, whole beans, hot dog slices, hard candies and nuts because they can easily choke on these foods.
5. Iron-deficiency anemia; children should receive 5.5 mg of iron per 1,000 kcalories. Enough milk must be provided to ensure adequate calcium and riboflavin intake but no more. After age 2, low-fat milk should be used so there will be more calories available for iron-rich foods such as lean meats, fish, poultry, eggs and legumes.
6. Nutrient deficiencies incur behavioral symptoms. Iron deficiency causes an energy crisis and directly affects moods, attention span, and learning ability. Television commercials promote sugary foods.
7. A true food allergy is an adverse reaction to foods that involves an immune response (common allergic reactions are skin rash, digestive upset, or respiratory discomfort); eggs, peanuts, and milk are likely to cause allergy. Food allergies can influence a person's nutrition status in that when trying to identify a suspected food, the food is omitted from the diet and this invites the risk of nutrient deficiencies.
8. Obese children are most likely to become obese adults and therefore are at risk for the social, economic and medical ramifications that often accompany obesity. Obese children begin puberty earlier and grow taller than their peers at first, but stop growing at a short height; they display higher levels of total cholesterol, triglycerides, LDL and VLDL; they tend to have high blood pressure; they are at risk for diabetes and asthma; they are victims of prejudice. Strategies include modifying dietary and exercise patterns. The dietary goal is to reduce the rate of weight gain; that is, to maintain weight while growing in height. Children are encouraged to eat slowly and select nutrient-dense foods and not be pressured to clean their plates. Daily physical activity should be included.
9. Parents can allow children to select from healthful choices, to prepare foods, to grow foods in a garden, and to visit food-related places.
10. School lunches can contribute to positive nutrition status if they are nutritious and acceptable. The provision of free or low-cost meals may positively impact the nutritional status of children from economically disadvantaged homes.
11. Nutrient needs increase during adolescence because it is a time of growth. Energy needs increase dramatically especially for males, iron needs increase especially for females because of monthly blood loss, and calcium needs increase for both genders.
12. Adolescents almost inevitably fall into irregular eating habits and snacks provide about a fourth of the average teenager's total daily food energy intake. Many snack foods offer a variety of nutrients although they are usually high in kcalories and low in vitamin A, folate, fiber and sometimes calcium.

13. Marijuana sometimes alters the sense of taste and increases enjoyableness of eating; prolonged use of the drug does not seem to increase energy intake enough to bring about weight gain. Cocaine causes lack of appetite and often addiction. Users often spend money on the drug instead of food, and lose interest in food during times they are using the drug. Alcohol is an empty-kcalorie beverage that can displace needed nutrients from the diet while simultaneously altering absorption and metabolism of nutrients. Tobacco smokers have lower intakes of dietary fiber, vitamin A, folate and vitamin C.

14. Smokers tend to have lower intakes of dietary fiber, vitamin A, beta-carotene, folate, and vitamin C. Since smokers require higher levels of some nutrients, their low intakes may increase their risks of cancer.

Sample Test Questions

1. c (p. 551)	9. d (p. 547, 558)	17. a (p. 569)	25. a (p. 572-573)
2. b (p. 552)	10. c (p. 562)	18. b (p. 576)	26. b (p. 571-572)
3. a (p. 550)	11. a (p. 558)	19. b (p. 574)	27. c (p. 578)
4. a (p. 551)	12. c (p. 562)	20. d (p. 567-569)	28. d (p. 579)
5. e (p. 551-552)	13. b (p. 568-569)	21. a (p. 571)	29. e (p. 579)
6. c (p. 555)	14. a (p. 566)	22. d (p. 577)	30. c (p. 579)
7. c (p. 555-556)	15. b (p. 564)	23. c (p. 575)	
8. b (p. 556)	16. c (p. 558-559)	24. e (p. 578)	

Short Answer Questions

1. Canned macaroni, canned spaghetti, cream of wheat, fortified dry cereals, noodles, rice or barley, tortillas, whole-wheat, enriched, or fortified bread, bran muffins, baked potato skins, cooked mushrooms, cooked mung bean sprouts or snow peas, green peas, mixed vegetable juice, canned plums, cooked dried apricots, dried peaches, raisins, bean dip, canned pork and beans, mild chili or other bean-meat dishes such as burritos, liverwurst, meat casseroles, peanut butter and jelly sandwich, lean roast beef or cooked ground beef, sloppy joes.

2. See Table 16-6.

Crossword Puzzle

1. lactadherin
2. alpha-lactalbumin
3. lactoferrin
4. *beikost*
5. wean
6. adolescence
7. food intolerances
8. colostrum
9. botulism
10. osteopenia

⑥ Chapter 17 – Life Cycle Nutrition: ⑥ Adulthood and the Later Years

Chapter Outline

I.   Nutrition and Longevity
    A. Observation of Older Adults
        1. Healthy Habits
        2. Physical Activity
    B. Manipulation of Diet
        1. Energy Restriction in Animals
        2. Energy Restriction in Human Beings
II. The Aging Process
    A. Physiological Changes
        1. Body Weight
        2. Body Composition
        3. Immune System
        4. GI Tract
        5. Tooth Loss
        6. Sensory Losses and Other Physical Problems
    B. Other Changes
        1. Psychological Changes
        2. Economic Changes
        3. Social Changes
III. Energy and Nutrient Needs of Older Adults
    A. Water
    B. Energy and Energy Nutrients
        1. Protein
        2. Carbohydrate and Fiber
        3. Fat
    C. Vitamins and Minerals
        1. Vitamin B_{12}
        2. Vitamin D
        3. Calcium
        4. Iron
    D. Nutrient Supplements

IV. Nutrition-Related Concerns of Older Adults
    A. Vision
        1. Cataracts
        2. Macular Degeneration
    B. Arthritis
        1. Osteoarthritis
        2. Rheumatoid Arthritis
        3. Gout
        4. Treatment
    C. The Aging Brain
        1. Nutrient Deficiencies and Brain Function
        2. Alzheimer's Disease
V.   Food Choices and Eating Habits of Older Adults
    A. Food Assistance Programs
    B. Meals for Singles
        1. Foodborne Illness
        2. Spend Wisely
        3. Be Creative
VI. Nutrient-Drug Interactions
    A. The Actions of Drugs
    B. The Interactions between Drugs and Nutrients
        1. Altered Food Intake
        2. Altered Nutrient Absorption
        3. Altered Drug Absorption
        4. Altered Metabolism
        5. Altered Nutrient Excretion
        6. Altered Drug Excretion
    C. The Inactive Ingredients in Drugs
        1. Sugar, Sorbitol, and Lactose
        2. Sodium

Summing Up

Adequate nutrition can help an individual reach the maximum life span by postponing and slowing (1)_____. Nutrition affects aging by its role in disease prevention. Six healthy habits to profoundly affect aging: abstinence from, or moderation in, (2)_____ use; regularity of meals; (3)_____ control; regular, adequate (4)_____; abstinence from (5)_____; and regular (6)_____. Some physiological changes that occur with age include: loss of (7)_____ and lean body mass; decline in (8)_____ system function; slow GI motility; and (9)_____ loss.

Setting nutrient standards for older people is difficult because (10)_____ differences become more pronounced as people age, and (11)_____ needs decline with advancing age. Other nutrient needs do not vary significantly with those of adults. However, elderly people often have problems obtaining the (12)_____ they need. Nutrition may provide some protection against some of the conditions associated with aging including cataracts and (13)_____. Food assistance programs provide (14)_____ meals to older adults.

Chapter Study Questions

1. What roles does nutrition play in aging, and what roles can it play in retarding aging?

2. What are some of the physiological changes that occur in the body's systems with aging? To what extent can aging be prevented?

3. Why does the risk of dehydration increase as people age?

4. Why do energy needs usually decline with advancing age?

5. Which vitamins and minerals need special consideration for the elderly? Explain why. Name some factors that complicate the task of setting nutrient standards for older adults.

6. Discuss the relationships between nutrition and cataracts and between nutrition and arthritis.

7. What characteristics contribute to malnutrition in older people?

Chapter Glossary

- **Alzheimer's disease:** a degenerative disease of the brain involving memory loss and major structural changes in neuron networks; also known as *senile dementia of the Alzheimer's type (SDAT)*, *primary degenerative dementia of senile onset*, or *chronic brain syndrome*.
- **arthritis:** inflammation of a joint, usually accompanied by pain, swelling, and structural changes.
- **atrophic gastritis:** a condition characterized by an inflamed stomach, bacterial overgrowth, and a lack of hydrochloric acid and intrinsic factor, all of which can impair the digestion and absorption of nutrients, most notably vitamin B_{12}, but also biotin, folate, calcium, iron, and zinc.
- **cataracts:** thickenings of the eye lenses that impair vision and can lead to blindness.
- **chronological age:** a person's age in years from his or her date of birth.
- **congregate meals:** nutrition programs that provide food for the elderly in conveniently located settings such as community centers.
- **dysphagia:** difficulty in swallowing.
- **edentulous:** the medical term for lack of teeth.
- **gout:** a common form of arthritis characterized by deposits of uric acid crystals in the joints.
- **life expectancy:** the average number of years lived by people in a given society.
- **life span:** the maximum number of years of life attainable by a member of a species.
- **longevity:** long duration of life.
- **macula:** a small, oval, yellowish region in the center of the retina that provides the sharp, straight-ahead vision so critical to reading and driving.
- **macular degeneration:** deterioration of the macular area of the eye that can lead to loss of central vision and eventual blindness.
- **Meals on Wheels:** a nutrition program that delivers food for the elderly to their homes.
- **neurofibrillary tangles:** snarls of the threadlike strands that extend from the nerve cells, commonly found in the brains of people with Alzheimer's dementia.
- **neurons:** nerve cells; the structural and functional units of the nervous system. Neurons initiate and conduct nerve impulse transmissions.
- **osteoarthritis:** a painful, degenerative disease of the joints that occurs when the cushioning cartilage in a joint deteriorates; joint structure is damaged, with loss of function; also called *degenerative arthritis*.
- **physiological age:** a person's age as estimated from her or his body's health and probable life expectancy.
- **pressure ulcers:** damage to the skin and underlying tissues as a result of compression and poor circulation; commonly seen in people who are bedridden or chairbound.
- **purines:** compounds of nitrogen-containing bases such as adenine, guanine, and caffeine. Purines that originate from the body are *endogenous* and those that derive from foods, *exogenous*.
- **quality of life:** a person's perceived physical and mental well-being.
- **rheumatoid arthritis:** a disease of the immune system involving painful inflammation of the joints and related structures.
- **sarcopenia:** loss of skeletal muscle mass, strength, and quality.
- **senile dementia:** the loss of brain function beyond the normal loss of physical adeptness and memory that occurs with aging.
- **senile plaques:** clumps of beta-amyloid on the nerve cells, commonly found in the brains of people with Alzheimer's dementia.
- **stress:** any threat to a person's well-being; a demand placed on the body to adapt.
- **stress response:** the body's response to stress, mediated by both nerves and hormones.
- **stressors:** environmental elements, physical or psychological, that cause stress.
- **ultrahigh temperature (UHT):** a process in which a food is exposed to temperatures above those of pasteurization just long enough to sterilize the food. Boxes of milk that can be stored at room temperature have been preserved through this process.

Sample Test Questions

Select the best answer for each question.

1. The maximum number of years of life attainable by a member of a species is called:
   a. life expectancy.
   b. life span.
   c. longevity.
   d. quality of life.

2. A person's age as estimated from his or her body's health and probable life expectancy is called:
   a. physiological age.
   b. chronological age.
   c. years of healthy life.
   d. longevity.

3. What is the life expectancy for black males and females in the U.S.?
   a. 68, 75 years
   b. 75, 80 years
   c. 79, 84 years
   d. 85, 89 years

4. What is the life expectancy for white males and females in the U.S.?
   a. 65, 70 years
   b. 75, 80 years
   c. 79, 84 years
   d. 85, 89 years

5. Studies of adults show that longevity is related, in part, to all of the following except:
   a. weight control.
   b. regularity of meals.
   c. short periods of sleep.
   d. no or moderate alcohol intake.

6. What would be the physiological age of a 75-year-old woman whose physical health is equivalent to that of her 50-year-old daughter?
   a. 25 years
   b. 50 years
   c. 70 years
   d. 125 years

7. A person's perceived physical and mental well being is called:
   a. longevity.
   b. life expectancy.
   c. life span.
   d. quality of life.

8. Environmental elements that cause the body to adapt are:
   a. stress responses.
   b. stresses.
   c. stressors.
   d. eustress.

9. Which of the following has been associated with regular physical activity in older adults?
   a. retention of sodium
   b. increase in blood LDL
   c. increased death rates
   d. slowing of cardiovascular aging

10. All of the following environmental factors are known to promote aging except:
    a. exercise.
    b. diseases.
    c. lack of nutrients.
    d. extremes of heat and cold.

11. Which of the following statements is true regarding exercise and older people?
    a. Regular physical activity provides a minimal benefit to older people.
    b. Older people should not engage in strength training.
    c. Physical activity may help older people maintain their independence.
    d. Loss of muscle mass is an inevitable part of aging.

12. Which of the following statements is true regarding energy restriction and aging?
    a. Energy restriction extends human life and should be encouraged.
    b. Energy restriction is beneficial for older people as long as it is extreme.
    c. The benefit of energy restriction for humans is unknown.
    d. Fasting for several days consecutively appears to be beneficial.

13. As people age, they often experience:
    a. declining immune system function.
    b. loss of elasticity of intestinal walls.
    c. tooth loss.
    d. all of the above
    e. a and b

14. Loss of skeletal muscle mass, strength and quality that often occurs with aging is called:
    a. dysphagia.
    b. hypertrophy.
    c. sarcopenia.
    d. sarcoma.

15. Dysphagia, which occurs among the elderly, is described as:
    a. difficulty chewing.
    b. difficulty swallowing.
    c. difficulty digesting.
    d. difficulty eliminating.

16. Atrophic gastritis is a condition that can especially impair the absorption of the following nutrients:
    a. vitamin B_{12} and biotin.
    b. calcium and iron.
    c. glucose and amino acids.
    d. all of the above
    e. a and b

17. Studies of the eating habits of older adults demonstrate all of the following except:
    a. those who live alone in federally funded housing had higher-quality diets.
    b. adults living alone consume insufficient amounts of food.
    c. malnutrition was associated with a lower level of education.
    d. men living with spouses ate higher quality diets than men living alone.

18. Nutrient needs of older people:
    a. vary according to individuals and show pronounced differences.
    b. increase; therefore, supplementation is required.
    c. remain the same as in young adult life.
    d. decrease.

19. Which of the following is a feature of elderly people and water metabolism?
    a. They do not feel thirsty or recognize dryness of the mouth.
    b. They have a higher total body water content compared with younger adults.
    c. They show increased frequency of urination which results in higher requirements.
    d. They frequently show symptoms of overhydration such as mental lapses and disorientation.

20. Because energy needs decrease, protein must be obtained from:
    a. liquid nutritional formulas.
    b. only vegan diets.
    c. fried chicken with gravy.
    d. low-kcalorie sources of high-quality protein.

21. To avoid constipation, older adults should:
    a. eat high-fiber foods and drink plenty of water.
    b. consume refined carbohydrate-containing foods.
    c. stop taking prescribed medications.
    d. eat only raw foods.

22. Which of the following statements describes one aspect of mineral nutrition of older adults?
    a. Zinc intake is adequate for about 95% of this group.
    b. Calcium intakes of females are near the RDA for this group.
    c. Iron-deficiency anemia in this population group is less common than in younger adults.
    d. Calcium allowances for this group have recently been increased by the Committee on Dietary Allowances.

23. What are the thickenings that occur to the lenses of the eye, thereby affecting vision, especially in the elderly?
    a. keratoids   c. retinitis
    b. cataracts   d. rhodopolids

24. What nutrients may be protective against cataract formation?
    a. iron and calcium   c. vitamin B_{12} and folate
    b. chromium and zinc   d. vitamin C and vitamin E

25. What nutrient(s) may be protective against macular degeneration?
    a. calcium   d. a and b
    b. antioxidant nutrients   e. b and c
    c. lutein

26. A connection exists between osteoarthritis and:
    a. calcium.   c. omega-3 fatty acids.
    b. vitamin D.   d. overweight.

27. How is nutrition linked to rheumatoid arthritis?
    a. The immune system relies on adequate nutrition.
    b. Omega-3 fatty acids may interfere with the action of prostaglandins.
    c. a and b
    d. none of the above

28. Clumps of beta-amyloid on the nerve cells are called:
    a. gout.   c. neurofibrillary tangles.
    b. senile plaques.   d. neurons.

29. Goals of the federal Nutrition Program for the elderly include the provision of all of the following except:
    a. transportation services.   c. opportunity for social interaction.
    b. high-cost nutritious meals.   d. counseling and referral to other social services.

30. To avoid foodborne illness, older people should:
    a. only eat raw foods.
    b. not eat or drink unpasteurized milk or milk products.
    c. not eat undercooked eggs, meat, poultry or fish.
    d. a and b
    e. b and c

Short Answer Questions

1. List 7 strategies for growing old healthfully.

    a.

    b.

    c.

    d.

    e.

f.

g.

Crossword Puzzle

Complete this crossword puzzle by Mary A. Wyandt, Ph.D., CHES.

Across:	Down:
3. any threat to a person's well-being; a demand placed on the body to adapt	1. nutrition programs that provide food for the elderly in a conveniently located setting such as a community center
7. chronic inflammation of the stomach accompanied by a diminished size and functioning of the mucosa and glands	2. a chemical agent released by one neuron that acts upon a second neuron or upon a muscle or gland cell and alters its electrical state or activity
8. environmental elements, physical or psychological, that cause stress	4. lack of teeth
9. long duration of life	5. nerve cells; the structural and functional units of the nervous system
10. loss of skeletal muscle mass, strength, and quality	6. difficulty in swallowing

⑤ Chapter 17 Answer Key ⑥

Summing Up

1. diseases
2. alcohol
3. weight
4. sleep
5. smoking
6. physical activity
7. bone
8. immune
9. tooth
10. individual
11. energy
12. nutrients
13. arthritis
14. nutritious

Chapter Study Questions

1. Nutrition can slow some aspects of the aging process within the natural limits set by heredity. Eating well can help prevent diseases common in old age (diabetes, obesity and CVD). Eating well can prevent deficiency diseases. Energy restriction in animals seems to lengthen their lives. Whether such dietary intervention is beneficial in human beings remains unknown.

2. Energy needs decrease; and older people tend to gain body fat and lose muscle mass. Hormone activity alters body composition, immune system changes raise the risk of infections, atrophic gastritis interferes with digestion and absorption, and tooth loss limits food choices. Depression, economics, social changes and loneliness contribute to poor food intake. Aging cannot be prevented.

3. Risk of dehydration increases as people age because older people do not feel thirsty or notice mouth dryness.

4. Energy needs usually decline with advancing age because lean body mass diminishes, reducing basal metabolic rate. As people age they may reduce their physical activity.

5. Some vitamin and mineral needs remain constant from early adulthood to later years; however, the nutrition needs of people 50 to 70 differ from those over 70. Individual differences become more pronounced as people grow older. The presence of chronic diseases and various medications influences nutrient needs. Dehydration is a risk for older adults. Energy needs decrease. Older adults must pay special attention to vitamin B_{12}, vitamin D, calcium and iron. Vitamin D deficiency is possible because most adults drink little or no milk. Iron-deficiency anemia occurs in many older adults. Zinc intake remains low in this population.

6. Oxidative stress plays a role in cataract development and antioxidant nutrients may help minimize the damage; consuming adequate intakes of vitamins C and E and carotenoids is important. Being overweight aggravates osteoarthritis partly because weight-bearing joints have to carry excess poundage. Weight loss may relieve some of this pain. Rheumatoid arthritis has possible links through the immune system. Some individuals may moderate the inflammatory response and experience some relief by consuming vegetables and olive oil.

7. Physically older people may not be mobile and as able to purchase and prepare nutritious meals. Psychologically, many older people live alone and may not prepare food for one; financially, many live on social security and have a limited income.

Sample Test Questions

1. b (p. 593)
2. a (p. 595)
3. a (p. 593)
4. b (p. 593)
5. c (p. 595)
6. b (p. 595)
7. d (p. 594)
8. c (p. 597)
9. d (p. 595)
10. a (p. 595)
11. c (p. 595)
12. c (p. 597)
13. d (p. 599)
14. c (p. 598)
15. b (p. 599)
16. e (p. 599)
17. a (p. 600)
18. a (p. 601)
19. a (p. 601)
20. d (p. 602)
21. a (p. 602)
22. c (p. 603)
23. b (p. 604)
24. d (p. 605)
25. e (p. 605)
26. d (p. 605)
27. a (p. 605)
28. b (p. 607)
29. b (p. 608)
30. e (p. 610)

Short Answer Questions

1. See Table 17-4 in the text.

Crossword Puzzle

1. congregate meals
2. neurotransmitter
3. stress
4. edentulous
5. neurons
6. dysphagia
7. atrophic gastritis
8. stressors
9. longevity
10. sarcopen

⑥ Chapter 18 – Diet and Health ⑨

Chapter Outline

I. Nutrition and Infectious Diseases
    A. The Immune System
        1. Phagocytes
        2. Lymphocytes: B-cells
        3. Lymphocytes: T-cells
    B. Nutrition and Immunity
    C. HIV and AIDS
II. Nutrition and Chronic Diseases
III. Cardiovascular Disease
    A. How Atherosclerosis Develops
        1. Causes of Atherosclerosis
        2. Plaques
        3. Blood Clots
        4. Blood Pressure and Atherosclerosis
        5. The Result: Heart Attacks and Strokes
    B. Risk Factors for Coronary Heart Disease
        1. Age, Gender, and Family History
        2. High LDL and Low HDL Cholesterol
        3. High Blood Pressure (Hypertension)
        4. Diabetes
        5. Obesity and Physical Inactivity
        6. Cigarette Smoking
        7. Atherogenic Diet
        8. Other Risk Factors
        9. Metabolic Syndrome
    C. Recommendations for Reducing Coronary Heart Disease Risk
        1. Cholesterol Screening
        2. Lifestyle Changes
IV. Hypertension
    A. How Hypertension Develops
    B. Risk Factors for Hypertension
    C. Treatment of Hypertension
        1. Weight Control
        2. Physical Activity
        3. The DASH Diet
        4. Drug Therapy

V. Diabetes Mellitus
    A. How Diabetes Develops
        1. Type 1 Diabetes
        2. Type 2 Diabetes
    B. Complications of Diabetes
        1. Diseases of the Large Blood Vessels
        2. Diseases of the Small Blood Vessels
        3. Diseases of the Nerves
    C. Recommendations for Diabetes
        1. Carbohydrate Sources
        2. Dietary Fat
        3. Protein
        4. Recommendations for Type 1 Diabetes
        5. Recommendations for Type 2 Diabetes
VI. Cancer
    A. How Cancer Develops
        1. Environmental Factors
        2. Dietary Factors—Cancer Initiators
        3. Dietary Factors—Cancer Promoters
        4. Dietary Factors—Antipromoters
VII. Recommendations for Chronic Diseases
    A. Recommendations for the Population
    B. Recommendations for Individuals
    C. Recommendations for Each Individual
VIII. Complementary and Alternative Medicine
    A. Defining Complementary and Alternative Medicine
    B. Sound Research, Loud Controversy
        1. Placebo Effect
        2. Risk versus Benefits
    C. Nutrition-Related Alternative Therapies
        1. Foods
        2. Vitamin and Mineral Supplements
        3. Herbal Remedies
        4. Herbal Precautions
    D. Internet Precautions
    E. The Consumer's Perspective

Summing Up

Nutrition cannot prevent or cure infectious diseases, but adequate intakes of all the nutrients can help support the (1)_____ system. If the immune system is impaired because of malnutrition or diseases such as AIDS, a person becomes vulnerable to (2)_____ diseases.

Due to the control of (3)_____ diseases, the average life expectancy today is longer, and the diseases most people now fear (other than AIDS) are: diseases of the (4)_____ and blood vessels, (5)_____, diabetes, lung diseases, and liver disease. Of the 10 leading causes of illness and death, 4 are directly associated with (6)_____. The risk factors for these degenerative diseases of adulthood are: (7)_____, behavioral, social, and (8)_____.

(9)_____ is a risk factor which aggravates the risk of almost every other disease. Food

(10)_____ are interwoven with these risk factors. Eating high- (11)_____ foods

and becoming (12)_____ increases the probabilities of contracting cancer,

(13)_____, diabetes, (14)_____ and diverticulosis.

      Lifestyle recommendations include: reduce consumption of fat and (15)_____, achieve

and maintain a desirable body (16)_____, increase consumption of complex carbohydrates and

(17)_____, reduce intake of (18)_____, and use (19)_____

in moderation, if at all.

      Major causes of death in the U.S. include diseases of the heart and (20)_____ vessels.

CVD accounts for more than (21)_____ of the nation's deaths per year, mostly by way of heart

attacks and strokes. (22)_____, high blood pressure and high blood cholesterol are the 3 major

risk factors for CVD. The big diet-related risk factors for CVD are glucose intolerance,

(23)_____, hypertension, and high blood cholesterol. Weight control, diet and

(24)_____ are all key factors in controlling these chronic diseases.

      The two types of diabetes mellitus are (25)_____ and type 2, which is most common.

People with type 1 diabetes coordinate diet, (26)_____ injections, and physical activity. People

with type 2 benefit most from a diet and physical activity program that promote (27)_____ loss.

      A high dietary fat intake is thought to increase chances of developing (28)_____. Other

factors related to cancer are: genetics, smoking, and water and air pollution. Food (29)_____

probably have little to do with the causation of cancer. Dietary recommendations are to decrease intake of certain

types of fats and increase plant (30)_____ intake. (31)_____ and tobacco use

increase the risk of cancer. Nonnutrient compounds in the (32)_____ vegetables offer a protective

effect against carcinogens. To prevent chronic diseases, recommendations focus on weight control and urge people

to limit saturated and (33)_____ fat, increase fiber-rich carbohydrates, and balance food intake

with (34)_____.

Chapter Study Questions

1. How do the major diseases of today as a group differ from those of several decades ago as a group? Why is nutrition considered so important in connection with today's major diseases?

2. What is HIV infection? What are the consequences of HIV infection?

3. Discuss the relationship between nutrition and chronic diseases.

4. Identify the major diet-related risk factors for atherosclerosis, hypertension, cancer, and diabetes.

5. Describe some ways in which people can alter their diets to lower their blood cholesterol levels.

6. Describe some steps that people with hypertension can take to lower their blood pressure.

7. Name the two major types of diabetes and describe some differences between them. How do dietary recommendations for each type of diabetes compare with the healthy diet recommended for all people?

8. Differentiate between cancer initiators, promoters, and antipromoters. Which nutrients or foods fit into each of these categories?

9. Describe the characteristics of a diet that might offer the best protection against the onset of cancer.

Chapter Glossary

- **adenomas:** cancers that arise from glandular tissues.
- **AIDS (acquired immune deficiency syndrome):** the late stage of HIV infection, in which severe complications develop.
- **angina:** a painful feeling of tightness or pressure in and around the heart, often radiating to the back, neck, and arms; caused by a lack of oxygen to an area of heart muscle.
- **antibodies:** large proteins of the blood and body fluids, produced by the immune system in response to the invasion of the body by foreign molecules (usually proteins called *antigens*). Antibodies combine with and inactivate the foreign invaders, thus protecting the body.

- **antigens:** substances that elicit the formation of antibodies or an inflammation reaction from the immune system.
- **antipromoters:** factors that oppose the development of cancer.
- **atheromatous plaque:** plaque associated with atherosclerosis.
- **atherosclerosis:** a type of artery disease characterized by plaques along the inner walls of the arteries.
- **autoimmune disorder:** a condition in which the body develops antibodies to its own proteins and then proceeds to destroy cells containing these proteins. In type 1 diabetes, the body develops antibodies to its insulin and destroys the pancreatic cells that produce the insulin, creating an insulin deficiency.
- **B-cells:** lymphocytes that produce antibodies. *B* stands for *bone marrow* where the B-cells develop and mature.
- **bioterrorism:** the intentional spreading of disease-causing microorganisms or toxins.
- **cancers:** malignant growths or tumors that result from abnormal and uncontrolled cell division.
- **carcinogenesis:** the process of cancer development.
- **carcinogens:** substances that can cause cancer (the adjective is *carcinogenic*).
- **carcinomas:** cancers that arise from epithelial tissues.
- **cardiovascular disease (CVD):** a general term of all diseases of the heart and blood vessels.
- **cerebral thrombosis:** a blood clot that blocks blood flow through an artery that feeds the brain.
- **CHD risk equivalents:** disorders that raise the risk of heart attacks, strokes, and other complications associated with cardiovascular disease to the same degree as existing CHD. These disorders include symptomatic carotid artery disease, peripheral arterial disease, abdominal aortic aneurysm, and diabetes mellitus.
- **complementary and alternative medicine (CAM):** diverse medical and health care systems, practices, and products that are not currently considered part of conventional medicine; also called *adjunctive, unconventional,* or *unorthodox therapies.*
- **complementary medicine:** an approach that uses alternative therapies as an adjunct to, and not simply a replacement for, conventional medicine.
- **conventional medicine:** diagnosis and treatment of diseases as practiced by medical doctors (M.D.) and doctors of osteopathy (D.O.) and allied health professionals such as physical therapists and registered nurses; also called *allopahty; Western, mainstream, orthodox,* or *regular medicine;* and *biomedicine.*
- **coronary heart disease (CHD):** the damage that occurs when the blood vessels carrying blood to the heart (the *coronary arteries*) become narrow and occluded.
- **coronary thrombosis:** a blood clot that blocks blood flow through an artery that feeds the heart muscle.
- **C-reactive protein (CRP):** a protein released during the acute phase of infection or inflammation that enhances immunity by promoting phagocytosis and activating platelets. Its presence may be used to assess a person's risk of an impending heart attack or stroke.
- **cytokines:** special proteins that direct immune and inflammatory responses.
- **diabetes mellitus:** a group of metabolic diseases characterized by hyperglycemia resulting from defects in insulin secretion, insulin action, or both.
- **embolism:** the obstruction of a blood vessel by an *embolus,* causing sudden tissue death.
- **embolus:** a traveling clot.
- **emerging risk factors:** recently identified factors that enhance the ability to predict disease risk in an individual.
- **gangrene:** the death of tissue, usually due to deficient blood supply.
- **genome:** the full complement of genetic material (DNA) in the chromosomes of a cell.
- **genomics:** the study of genomes.
- **gliomas:** cancers that arise from glial cells of the central nervous system.
- **heart attack:** sudden tissue death caused by blockages of vessels that feed the heart muscle; also called *myocardial infarction* or *cardiac arrest.*
- **HIV (human immunodeficiency virus):** the virus that causes AIDS. The infection progresses to become an immune system disorder that leaves its victims defenseless against numerous infections.
- **hyperglycemia:** elevated blood glucose concentrations.
- **hypertension:** higher-than-normal blood pressure (BP = $\geq$140/$\geq$90). Hypertension that develops without an identifiable cause is known as *essential* or *primary hypertension*; hypertension that is caused by a specific disorder such as kidney disease is known as *secondary hypertension.*

- **immune system:** the body's natural defense against foreign materials that have penetrated the skin or mucous membranes.
- **immunoglobulins:** proteins capable of acting as antibodies.
- **inflammation:** an immunological response to cellular injury characterized by an increase in white blood cells, redness, heat, pain, swelling and often loss of function of the affected body part.
- **initiators:** factors that cause mutations that give rise to cancer, such as radiation and carcinogens.
- **insulin resistance:** the condition in which a normal amount of insulin produces a subnormal effect, resulting in an elevated fasting glucose; a metabolic consequence of obesity that precedes type 2 diabetes.
- **integrative medicine:** an approach that incorporates alternative therapies into the practice of conventional medicine (similar to complementary medicine, but a closer relationship is implied).
- **leukemias:** cancers that arise from the white blood cell precursors.
- **lymphocytes:** white blood cells that participate in acquired immunity; B-cells and T-cells.
- **lymphomas:** cancers that arise from lymph tissue.
- **macrophages:** large, phagocytic cells of the immune system.
- **malignant:** describes a cancerous cell or tumor, which can injure healthy tissue and spread cancer to other regions of the body.
- **melanomas:** cancers that arise from pigmented skin cells.
- **metabolic syndrome:** a combination of risk factors—insulin resistance, hypertension, abnormal blood lipids, and abdominal obesity—that greatly increase a person's risk of developing coronary heart disease; also called *Syndrome X, insulin resistance syndrome,* or *dysmetabolic syndrome.*
- **metastasize:** the spread of cancer from one part of the body to another.
- **microangiopathies:** disorders of the small blood vessels.
- **peripheral resistance:** the resistance to pumped blood in the small arterial branches (arterioles) that carry blood to tissues.
- **phagocytes:** white blood cells (neutrophils and macrophages) that have the ability to ingest and destroy foreign substances.
- **phagocytosis:** the process by which phagocytes engulf and destroy foreign materials.
- **plaques:** mounds of lipid material, mixed with smooth muscle cells and calcium, that develop in the artery walls in atherosclerosis.
- **platelets:** tiny, disc-shaped bodies in the blood, important in blood clot formation.
- **prediabetes:** condition in which blood glucose levels are higher than normal but not high enough to be diagnosed as diabetes; considered a major risk factor for future diabetes and cardiovascular diseases. Formerly called *impaired glucose tolerance.*
- **prehypertension:** slightly higher-than-normal blood pressure, but not as high as hypertension (BP = 120-139/80-89).
- **promoters:** factors that favor the development of cancers once they have begun.
- **sarcomas:** cancers that arise from connective tissues, such as muscle or bone.
- **stroke:** an event in which the blood flow to a part of the brain is cut off; also called *cerebrovascular accident (CVA).*
- **synergistic:** multiple factors operating together in such a way that their combined effects are greater than the sum of their individual effects.
- **T-cells**: lymphocytes that attack antigens. *T* stands for the thymus gland, where the T-cells mature.
- **thrombosis:** the formation of a *thrombus.*
- **thrombus:** a blood clot that may obstruct a blood vessel, causing gradual tissue death.
- **transient ischemic attack (TIA):** a temporary reduction in blood flow to the brain, which causes temporary symptoms that vary depending of the part of the brain affected. Common symptoms include light-headedness, visual disturbances, paralysis, staggering, numbness, and inability to swallow.
- **tumor:** an abnormal tissue mass with no physiological function; also called a *neoplasm.*
- **type 1 diabetes:** the type of diabetes that accounts for 5 to 50% of diabetes cases and usually results from autoimmune destruction of pancreatic beta cells. In this type of diabetes, the pancreas produces little or no insulin.
- **type 2 diabetes:** the type of diabetes that accounts for 90 to 95% of diabetes cases and usually results from insulin resistance coupled with insufficient insulin secretion. Obesity is present in 80 to 90% of cases.

Sample Test Questions

Select the best answer for each question.

1. The diseases most feared today are:
   1. tuberculosis.
   2. smallpox.
   3. diseases of the heart and blood vessels.
   4. cancer.
   5. diabetes.

   a. 1, 2, 3
   b. 2, 3, 4
   c. 3, 4, 5
   d. 1, 3, 5

2. Which disease has some relationship with nutrition?
   a. cancer
   b. diabetes
   c. heart disease
   d. a and b
   e. a, b, and c

3. What body system usually defends the body against infectious diseases?
   a. synergistic system
   b. inflammation
   c. immune system
   d. autoimmune system

4. The intentional spreading of toxins is called:
   a. bioterrorism.
   b. terrorism.
   c. violence.
   d. autoinoculation.

5. Organs of the immune system include the:
   a. heart.
   b. spleen.
   c. thymus.
   d. a and b
   e. b and c

6. Cells of the immune system include:
   a. phagocytes.
   b. lymphocytes.
   c. B-cells.
   d. a and c
   e. a, b, and c

7. What proteins are secreted by phagocytes to activate the immune response?
   a. neutrophils
   b. cytokines
   c. macrophages
   d. T-cells

8. What type of cells produce antibodies in response to infection?
   a. B-cells
   b. T-cells
   c. platelets
   d. immunoglobulins

9. Nutrients known to affect immunity include:
   a. protein, folate, and fatty acids.
   b. carbohydrate, protein, and zinc.
   c. calcium, vitamin A, and selenium.
   d. B vitamins, carbohydrate, and vitamin E.

10. Disease and malnutrition create a _____ downward spiral.
    a. resistant
    b. synergistic
    c. antagonistic
    d. circular

11. Which of the following statements is true regarding HIV/AIDS?
    a. AIDS prevention depends on good nutrition.
    b. A cure has been developed for HIV/AIDS.
    c. HIV is transmitted by direct contact with contaminated body fluids.
    d. Condoms are 100% effective at preventing HIV transmission.

12. Being obese increases the probabilities of contracting which of the following?
    a. cancer
    b. hypertension
    c. diabetes
    d. atherosclerosis
    e. all of the above

13. Mounds of lipid material mixed up with smooth muscle cells and calcium which develop in the artery walls are called:
    a. osteoporosis.
    b. atherosclerosis.
    c. CVD.
    d. plaques.

14. The immunological response to tissue damage involved with atherosclerosis is called:
    a. hyperglycemia.
    b. hypoglycemia.
    c. inflammation.
    d. irritation.

15. Which of the following is released during the acute phase of inflammation and activates platelets?
    a. C-reactive protein
    b. plaques
    c. cholesterol
    d. fatty liver

16. The event in which an embolus lodges in vessels that feed the heart muscle, causing sudden tissue death, is called a:
    a. thrombus.
    b. heart attack.
    c. stroke.
    d. aorta.

17. Damage that occurs when the blood vessels carrying blood to the heart become narrow and occluded is called:
    a. an aneurysm.
    b. a plaque.
    c. an aorta.
    d. coronary heart disease.

18. Which two conditions worsen each other?
    a. cancer, diverticulosis
    b. cancer, diabetes
    c. hypertension, atherosclerosis
    d. hypoglycemia, hypertension

19. Risk factors for atherosclerosis that can be minimized by behavior change include:
    1. smoking.
    2. hypertension.
    3. gender.
    4. lack of exercise.
    5. obesity.
    6. heredity.
    7. stress.

    a. 1, 2, 3, 4, 5
    b. 1, 2, 4, 5, 6
    c. 2, 3, 4, 5, 7
    d. 1, 2, 4, 5, 7

20. A condition in which the body develops antibodies to its own proteins and then destroys cells containing these proteins is called:
    a. an inflammation disorder.
    b. AIDS.
    c. hyperglycemia.
    d. an autoimmune disorder.

21. A cluster of risk factors including low HDL, high blood pressure, insulin resistance and abdominal obesity is known as:
    a. emerging risk factors.
    b. metabolic syndrome.
    c. peripheral resistance.
    d. an autoimmune disorder.

22. When the pancreas loses its ability to synthesize insulin, this is called:
   a. type 1 diabetes.
   b. type 2 diabetes.
   c. prediabetes.
   d. hyperlipidemia.

23. Diseases that result from the unchecked growth of malignant tumors are:
   a. AIDS.
   b. cancers.
   c. hypertension.
   d. diverticulosis.

24. A _____ diet is linked to cancer development.
   a. high-fat.
   b. low-fat.
   c. high-fiber.
   d. high-food additive.

25. What term indicates that cancer has spread to other parts of the body?
   a. malignant
   b. benign
   c. promoter
   d. metastasize

26. What term refers to dietary factors that start cancer development?
   a. promoters
   b. initiators
   c. antipromoters
   d. tumors

27. To inhibit cancer promotion,
   a. consume high-protein and high-fat diets.
   b. reduce saturated and *trans* fat intakes.
   c. increase omega-3 fatty acid intake.
   d. a and b
   e. b and c

28. Dietary antipromoters include:
   a. fruits and vegetables.
   b. grilled meats.
   c. refined breads.
   d. B vitamins.

29. Recommendations for reducing chronic disease risk include:
   a. take multivitamin/mineral supplements daily.
   b. increase fat intake.
   c. severely restrict kcalories.
   d. achieve and maintain a healthy body weight.

30. Health recommendations that urge dietary changes only for people who are known to need them are taking a:
   a. preventive approach.
   b. population approach.
   c. secondary approach.
   d. medical approach.

Short Answer Questions

1. The 4 steps in cancer development are:

   a.

   b.

   c.

   d.

2. List the five possible criteria for metabolic syndrome:

   a.

   b.

   c.

   d.

   e.

Crossword Puzzle

Complete this crossword puzzle by Mary A. Wyandt, Ph.D., CHES.

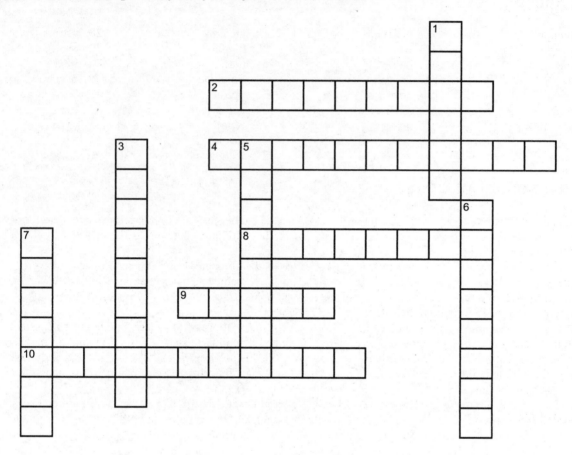

Across:	Down:
2. tumors that multiply out of control, threaten health, and require treatment	1. tumors that stop growing without intervention or can be removed surgically and pose no threat to health
4. substances or agents that are capable of causing cancer	3. cancers that arise from lymph tissue
8. a new growth of tissue forming an abnormal mass with no function	5. cancers that arise from glandular tissue
9. another word for neoplasm; those that pose no problem are called benign; those that resist treatment and are harmful are malignant	6. the obstruction of a blood vessel by an embolus
10. to spread from one part of the body to another	7. cancers that arise from glial cells of the central nervous system

⊙ Chapter 18 Answer Key ⊚

Summing Up

1. immune
2. infectious
3. infectious
4. heart
5. cancer
6. nutrition
7. environmental
8. genetic
9. Obesity
10. behaviors
11. fat
12. obese
13. hypertension
14. atherosclerosis
15. cholesterol
16. weight
17. fiber
18. sodium
19. alcohol
20. blood
21. half
22. Smoking
23. obesity
24. exercise
25. type 1
26. insulin
27. weight
28. cancer
29. additives
30. fiber
31. Alcohol
32. cruciferous
33. *trans*
34. exercise

Chapter Study Questions

1. The major diseases of today are related to lifestyle, whereas the diseases of several decades ago were mostly infectious. Nutrition is considered so important in connection with today's major diseases because it is involved with risk factors.

2. HIV infection is human immunodeficiency virus; it attacks the immune system and causes those who are infected to be vulnerable to opportunistic infections. It is not curable and in its late stages it usually causes severe weight loss, tuberculosis, recurrent bacterial pneumonia, CNS infections, GI tract infections, skin infections, cancers, and severe diarrhea.

3. Four of the ten leading causes of death in the U. S. have some relationship with diet. Single nutrients affect single diseases such as calcium and osteoporosis, saturated fat and heart disease. Also each nutrient may have connections with several diseases because its role in the body is not specific to a disease, but to a body function. Attention to nutrition may prevent some chronic diseases, which may improve the quality of life and slow disease progression.

4. Risk factors are environmental, behavioral, social and genetic; specific risk factors are: high-fat diet, excessive alcohol consumption, high salt intake, contaminated food intake, low complex carbohydrate and fiber intake, low vitamin and/or mineral intakes, high sugar intake, low calcium intake, stress, sedentary lifestyle, smoking, and obesity.

5. Maintain appropriate body weight, exercise regularly, do not smoke, and control stress. To control blood cholesterol, consume an overall low-fat diet with less than 7% of total kcal from saturated fat, less than 1% of total kcal from *trans* fat, and less than 300 mg/day of cholesterol.

6. Balance energy intake and output to maintain appropriate body weight, lower salt intake, exercise, eat adequate amounts of calcium, lower saturated fat intake, and, if you drink alcohol, do so in moderation. Consume the DASH (Dietary Approaches to Stop Hypertension) diet, which is rich in fruits, vegetables, nuts, and low-fat milk products and is low in fat and saturated fat.

7. Two types of diabetes are type 1 and type 2. In type 1 (an autoimmune disorder) the person produces no insulin; type 2 is more common and is associated with obesity and insulin resistance. Nutrition therapy focuses on maintaining optimal nutrition status, controlling blood glucose, achieving a desirable lipid profile, controlling blood pressure and preventing and treating the complications of diabetes. Distributing carbohydrate intake throughout the day is a key strategy. Exchange lists are one method used to plan diets, and general dietary guidelines for good health are similar to meal planning for diabetes.

8. Initiators are carcinogens that intrude into cells and alter the genetic material. Promoters enhance cancer development; antipromoters offer protective effects. High levels of pesticides, acrylamide, and alcohol are initiators; fat and excessive kcalories are promoters; fruits, vegetables, and antioxidants are antipromoters.

9. Avoid obesity; reduce the consumption of total fat; eat more high-fiber foods, such as whole-grain cereals, vegetables, and fresh fruits; include a variety of vegetables and fruits in the daily diet; avoid possible carcinogens by limiting consumption of charbroiled or fried foods; and limit or stop consumption of alcoholic beverages.

Sample Test Questions

1. c (p. 624)	9. a (p. 623)	17. d (p. 626)	25. d (p. 642)
2. e (p. 624)	10. b (p. 623)	18. c (p. 629)	26. b (p. 643)
3. c (p. 622)	11. c (p. 623)	19. d (p. 628)	27. e (p. 645)
4. a (p. 621)	12. e (p. 625)	20. d (p. 638)	28. a (p. 645)
5. e (p. 622)	13. d (p. 626)	21. b (p. 630)	29. d (p. 646)
6. e (p. 622)	14. c (p. 626)	22. a (p. 638)	30. d (p. 647)
7. b (p. 622)	15. a (p. 627)	23. b (p. 642)	
8. a (p. 622)	16. b (p. 627)	24. a (p. 645)	

Short Answer Questions

1.
   a. exposure to a carcinogen
   b. entry of the carcinogen into a cell
   c. initiation, probably by the carcinogen's altering of the cellular DNA
   d. enhancement of cancer development by promoters, probably involving several more steps before the cell begins to multiply out of control; tumor formation
2.
   a. abdominal obesity
   b. triglycerides greater than or equal to 150 mg/dL
   c. HDL less than 40 mg/dL in men or less than 50 mg/dL in women
   d. blood pressure greater than or equal to 130/85 mm Hg
   e. fasting blood glucose greater than or equal to 100 mg/dL

Crossword Puzzle

1. benign	4. carcinogens	7. gliomas	10. metastasize
2. malignant	5. adenomas	8. neoplasm	
3. lymphomas	6. embolism	9. tumor	

⑥ Chapter 19 – Consumer Concerns ⑥
about Foods and Water

Chapter Outline

I. Foodborne Illnesses
   A. Foodborne Infections and Food Intoxications
      1. Foodborne Infections
      2. Food Intoxications
   B. Food Safety in the Marketplace
      1. Industry Controls
      2. Consumer Awareness
   C. Food Safety in the Kitchen
      1. Safe Handling of Meats and Poultry
      2. Mad Cow Disease
      3. Avian Influenza
      4. Safe Handling of Seafood
      5. Other Precautions and Procedures
   D. Food Safety While Traveling
   E. Advances in Food Safety
      1. Irradiation
      2. Consumer Concerns about Irradiation
      3. Regulation of Irradiation
      4. Other Pasteurizing Systems
II. Nutritional Adequacy of Food and Diets
   A. Obtaining Nutrient Information
   B. Minimizing Nutrient Losses
III. Environmental Contaminants
   A. Harmfulness of Environmental Contaminants
      1. Methylmercury
      2. PBB and PCB
   B. Guidelines for Consumers
III. Natural Toxicants in Foods
IV. Pesticides
   A. Hazards and Regulation of Pesticides
      1. Hazards of Pesticides
      2. Regulation of Pesticides
      3. Pesticides from Other Countries
   B. Monitoring Pesticides
      1. Food in the Fields
      2. Food on the Plate
   C. Consumer Concerns

      1. Minimizing Risks
      2. Alternatives to Pesticides
      3. Organically Grown Crops
V. Food Additives
   A. Regulations Governing Additives
      1. The GRAS List
      2. The Delaney Clause
      3. Margin of Safety
      4. Risks versus Benefits
   B. Intentional Food Additives
      1. Antimicrobial Agents
      2. Antioxidants
      3. Colors
      4. Artificial Flavors and Flavor Enhancers
      5. Texture and Stability
      6. Nutrient Additives
   C. Indirect Food Additives
      1. Acrylamide
      2. Microwave Packaging
      3. Dioxins
      4. Decaffeinated Coffee
      5. Hormones
      6. Antibiotics
VI. Consumer Concerns about Water
   A. Sources of Drinking Water
   B. Water Systems and Regulation
      1. Home Water Treatments
      2. Bottled Water
VII. Food Biotechnology
   A. The Promises of Genetic Engineering
      1. Extended Shelf Life
      2. Improved Nutrient Composition
      3. Efficient Food Processing
      4. Efficient Drug Delivery
      5. Genetically Assisted Agriculture
      6. Other Possibilities
   B. The Potential Problems and Concerns
   C. FDA Regulations

Summing Up

    Food safety concerns include (1)_____ illnesses, nutritional adequacy of foods, environmental contaminants, naturally occurring toxicants, (2)_____, food additives, and water safety. Foodborne illnesses can be caused by either an infection or an (3)_____. Food intoxications are caused by eating foods containing (4)_____ toxins or microbes that produce toxins. To prevent foodborne illnesses, keep a clean, safe kitchen, avoid (5)_____, keep hot foods hot, and keep (6)_____ foods cold. Meat and poultry should not be (7)_____ or rinsed. People should not eat (8)_____ seafood. Use of low-dose (9)_____ has helped control the safety of foods.

To minimize nutrient losses, consumers can refrigerate fruits and vegetables, (10)_____ fruits and vegetables before cutting them, store them in airtight containers, and cook them for (11)_____ times in (12)_____ water. Environmental contamination of foods is a concern. Consumers can remain (13)_____ to the possibility of contamination and listen for (14)_____ health announcements. Additionally, consumers can eat a (15)_____ of foods.

Pesticides are used to improve (16)_____ yields but may be hazardous if used (17)_____. The (18)_____ tests foods for pesticide residues. Consumers can take measures to reduce pesticide intake including buying (19)_____ produce. (20)_____ use has benefits and the FDA regulates the use of specific additives. Water may contain infectious microorganisms, (21)_____ contaminants, or (22)_____ residues. The (23)_____ monitors the safety of the public water system.

Chapter Study Questions

1. To what extent does food poisoning present a real hazard to consumers eating U.S. foods? How often does it occur?

2. Distinguish between the two types of foodborne illnesses and provide an example of each. Describe several measures that help prevent foodborne illnesses.

3. What special precautions apply to meats? To seafood?

4. What is meant by a "persistent" contaminant of foods? Describe how contaminants get into foods and build up in the food chain.

5. What dangers do natural toxicants present?

6. How do pesticides become a hazard to the food supply, and how are they monitored? In what ways can people reduce the concentrations of pesticides in and on foods that they prepare?

7. What is the difference between a GRAS substance and a regulated food additive? Give examples of each. Name and describe the different classes of additives.

Chapter Glossary

- **additives:** substances not normally consumed as foods but added to food either intentionally or by accident.
- **BHA** and **BHT:** preservatives commonly used to slow the development of off-flavors, odors, and color changes caused by oxidation.
- **bioaccumulation:** the accumulation of contaminants in the flesh of animals high on the food chain.
- **bovine growth hormone (BGH):** a hormone produced naturally in the pituitary gland of a cow that promotes growth and milk production; now produced for agricultural use by bacteria.
- **carcinogen:** a substance that causes cancer.
- **certification:** the process in which a private laboratory inspects shipments of a product for selected chemicals and then, if the product is free of violative levels of those chemicals, issues a guarantee to that effect.
- **contaminants:** substances that make a food impure and unsuitable for ingestion.
- **cross-contamination:** the contamination of food by bacteria that occurs when the food comes into contact with surfaces previously touched by raw meat, poultry, or seafood.
- **Delaney Clause:** a clause in the Food Additive Amendment to the Food, Drug, and Cosmetic Act that states that no substance that is known to cause cancer in animals or human beings at any dose level shall be added to foods.
- **dioxins:** a class of chemical pollutants created as by-products of chemical manufacturing, incineration, chlorine bleaching of paper pulp, and other industrial processes. Dioxins persist in the environment and accumulate in the food chain.
- **foodborne illness:** illness transmitted to human beings through food and water, caused by either an infectious agent (foodborne infection) or a poisonous substance (food intoxication); commonly known as *food poisoning*.
- **food chain:** the sequence in which living things depend on other living things for food.
- **generally recognized as safe (GRAS):** food additives that have long been in use and are believed safe. First established by the FDA in 1958, the GRAS list is subject to revision as new facts become known.
- **genotoxicant:** a substance that mutates or damages genetic material.
- **hazard:** a source of danger; used to refer to circumstances in which harm is possible under normal conditions of use.
- **Hazard Analysis Critical Control Points (HACCP):** a systematic plan to identify and correct potential microbial hazards in the manufacturing, distribution, and commercial use of food products; commonly referred to as "HASS-ip."
- **heavy metal:** any of a number of mineral ions such as mercury and lead, so called because they are of relatively high atomic weight. Many heavy metals are poisonous.

- **indirect** or **incidental additives:** substances that can get into food as a result of contact with foods during growing, processing, packaging, storing, cooking, or some other stage before the foods are consumed; sometimes called *accidental additives*.
- **intentional food additives:** additives intentionally added to foods, such as nutrients, colors, and preservatives.
- **irradiation:** sterilizing a food by exposure to energy waves, similar to ultraviolet light and microwaves.
- **margin of safety:** when speaking of food additives, a zone between the concentration normally used and that at which a hazard exists. For common table salt, for example, the margin of safety is $^1/_5$ (five times the amount normally used would be hazardous).
- **monosodium glutamate (MSG):** a sodium salt of the amino acid glutamic acid commonly used as a flavor enhancer. The FDA classifies MSG as a "generally recognized as safe" ingredient.
- **MSG symptom complex:** an acute, temporary intolerance reaction that may occur after the ingestion of the additive MSG (monosodium glutamate). Symptoms include burning sensations, chest and facial flushing and pain, and throbbing headaches.
- **nitrites:** salts added to food to prevent botulism. One example is sodium nitrite, which is used to preserve meats.
- **nitrosamines:** derivatives of nitrites that may be formed in the stomach when nitrites combine with amines. Nitrosamines are carcinogenic in animals.
- **organic:** in agriculture, crops grown and processed according to USDA regulations defining the use of fertilizers, herbicides, insecticides, fungicides, preservatives, and other chemical ingredients.
- **organic halogens:** an organic compound containing one or more atoms of a halogen—fluorine, chlorine, iodine, or bromine.
- **pasteurization:** heat processing of food that inactivates some, but not all, microorganisms in the food; not a sterilization process. Bacteria that cause spoilage are still present.
- pathogens: microorganisms capable of producing disease.
- **PBB (polybrominated biphenyl)** and **PCB (polychlorinated biphenyl):** toxic organic compounds used in pesticides, paints, and flame retardants.
- **persistence:** stubborn or enduring continuance; with respect to food contaminants, the quality of persisting, rather than breaking down, in the bodies of animals and human beings.
- **pesticides:** chemicals used to control insects, weeds, fungi, and other pests on plants, vegetables, fruits, and animals. Used broadly, the term includes herbicides (to kill weeds), insecticides (to kill insects), and fungicides (to kill fungi).
- **potable water:** water that is suitable for drinking.
- **preservatives:** antimicrobial agents, antioxidants, and other additives that retard spoilage or maintain desired qualities, such as softness in baked goods.
- **residues:** whatever remains. In the case of pesticides, those amounts that remain on or in foods when people buy and use them.
- **risk:** a measure of the probability and severity of harm.
- **safety:** the condition of being free from harm or danger.
- **solanine:** a poisonous narcotic-like substance present in potato peels and sprouts.
- **sulfites:** salts containing sulfur that are added to foods to prevent spoilage.
- **sushi:** vinegar-flavored rice and seafood, typically wrapped in seaweed and stuffed with colorful vegetables. Some sushi is stuffed with raw fish; other varieties contain cooked seafood.
- **tolerance level:** the maximum amount of a residue permitted in a food when a pesticide is used according to label directions.
- **toxicity:** the ability of a substance to harm living organisms. All substances are toxic if high enough concentrations are used.
- **travelers' diarrhea:** nausea, vomiting, and diarrhea caused by consuming food or water contaminated by any of several organisms, most commonly, *E. coli, Shigella, Campylobacter jejuni,* and *Salmonella.*
- **ultrahigh temperature (UHT) treatment:** sterilizing a food by brief exposure to temperatures above those normally used.

Sample Test Questions

Select the best answer for each question.

1. A term used to describe circumstances in which danger is possible under normal conditions of use is:
   a. *pesticide.*
   b. *risk.*
   c. *hazard.*
   d. *toxicity.*

2. The ability of a substance to harm living organisms is called:
   a. toxicity.
   b. a hazard.
   c. risk.
   d. irradiation.

3. A measure of the probability and severity of harm is called:
   a. toxicity.
   b. hazardous.
   c. un-safety.
   d. risk.

4. Consumers rely on monitoring agencies to set _____ standards.
   a. risk
   b. hazard
   c. safety
   d. benefit

5. First on the FDA's list of priority concerns related to food is:
   a. foodborne infection.
   b. environmental contaminants.
   c. pesticide residues.
   d. food additives.
   e. food toxicants.

6. Foodborne illnesses are:
   a. minor irritations.
   b. transmitted from one human being to another.
   c. caused by an infectious agent or a poisonous substance.
   d. a and b
   e. b and c

7. Microorganisms capable of causing disease are called:
   a. intoxicants.
   b. pathogens.
   c. residues.
   d. additives.

8. Possible symptoms of foodborne illness that require medical help include:
   a. bloody diarrhea.
   b. diarrhea lasting more than 3 days.
   c. difficulty breathing or swallowing.
   d. fever lasting more than 24 hours.
   e. all of the above

9. Common foodborne pathogens include:
   a. *Salmonella.*
   b. *Campylobacter jejuni.*
   c. solanine.
   d. a and b
   e. a, b and c

10. The most common food intoxicant is:
   a. *Staphylococcus aureus.*
   b. *Clostridium botulinum.*
   c. *Listeria.*
   d. *Escherichia coli.*

11. Pasteurization is heat process that:
   a. sterilizes food.
   b. causes exposure to energy waves.
   c. kills bacteria that cause spoilage.
   d. inactivates some microorganisms in food.

12. Which of the following statements is true regarding the Hazard Analysis Critical Control Points (HACCP)?
    a. It is a plan to control food pesticide use.
    b. It requires food manufacturers to implement controls to prevent foodborne disease.
    c. It is a plan designed to eliminate all use of ultrahigh temperature treatment.
    d. It is a plan to stop water fluoridation.

13. When cooked foods come in contact with surfaces touched by raw meat, this is called:
    a. HACCP.
    b. cross-contamination.
    c. irradiation.
    d. bioaccumulation.

14. To avoid foodborne illnesses at home:
    a. keep a clean, safe kitchen.
    b. avoid cross-contamination.
    c. keep hot foods hot.
    d. keep cold foods cold.
    e. all of the above

15. Which of the following should you do to avoid cross-contamination?
    a. Wash raw meat and poultry.
    b. Rinse cutting boards after cutting raw meat.
    c. Do not wash fruits and vegetables.
    d. Separate raw, cooked, and ready-to-eat foods.

16. To avoid mad cow disease, consumers can:
    a. select whole cuts of meat instead of ground beef or sausage.
    b. cook meat to at least 140 degrees.
    c. consume organs of meat instead of muscle.
    d. use pasteurization.

17. Which of the following statements is true?
    a. Bird flu may be transmitted by eating poultry.
    b. Sushi is dangerous to eat.
    c. Freezing fish can kill all worms and worm eggs.
    d. Eating raw oysters is dangerous.

18. Foods that are frequently unsafe include:
    a. peeled fruit, high-sugar foods, and steaming-hot foods.
    b. soft cheeses, salad bar items, and hamburgers.
    c. raw milk, raw sprouts and scallions, and undercooked eggs.
    d. sandwiches, unwashed berries, and grapes.

19. Nausea, vomiting, and other intestinal problems caused by consuming contaminated food or water are called:
    a. travelers' diarrhea.
    b. indigestion.
    c. dysentery.
    d. gas.

20. What treatment controls mold in grains, sterilizes spices and teas, and controls insects in fresh fruits and vegetables?
    a. irradiation
    b. pasteurization
    c. ultrahigh temperature treatment
    d. microwaving

21. The major source of contaminants, such as heavy metals, in the food chain is:
    a. industry.
    b. agriculture.
    c. nature.
    d. microorganisms.

22. Pesticide use is <u>not</u> monitored by the:
    a. USPS.
    b. FAO.
    e. b and c
    c. EPA.
    d. HACCP.
    f. a and d

23. To protect infants and children from pesticide poisoning, government agencies set a:
    a. persistence amount.
    b. tolerance level.
    c. certification contact.
    d. bioaccumulation level.

24. A poisonous, narcotic-like substance present in potato peels and sprouts is called:
    a. an organic halogen.
    b. solanine.
    c. polybrominated biphenyl.
    d. an incidental additive.

25. Food additives are the concern of the:
    a. GRAS.
    b. USDA.
    c. FDA.
    d. CFC.

26. Substances widely used for many years without apparent ill effects are on the _____ list.
    a. FDA
    b. GRAS
    c. Delaney
    d. Additive Safety

27. The Delaney Clause states that no additives known to cause cancer in:
    a. animals or people at any dose level can be used.
    b. people at any dose level can be used.
    c. animals or people at one-gram levels can be used.
    d. people at one-gram levels can be used.

28. Food additives may be used to:
    a. disguise faulty products.
    b. deceive customers.
    c. destroy nutrients.
    d. enhance flavor.

29. Antimicrobial agents protect foods against:
    a. oxidation.
    b. organisms.
    c. radiation.
    d. pesticides.

30. Incidental additives find their way into food as a result of:
    a. increasing nutritive value of foods.
    b. home cooking errors.
    c. manufacturing procedures.
    d. advertising gimmicks.

Crossword Puzzle

Complete this crossword puzzle by Mary A. Wyandt, Ph.D., CHES.

Across:	Down:
2. the process in which a private laboratory inspects shipments of a product for selected chemicals and then, if the product is free of volatile levels of those chemicals, issues a guarantee to that effect	1. stubborn or enduring continuance
	3. the maximum amount of a residue permitted in a food when a pesticide is used according to label directions
4. a poisonous narcotic-like substance present in potato peels and sprouts	6. the condition of being free from harm or danger
5. sterilizing a food by exposure to energy waves, similar to ultraviolet light and microwaves	8. source of danger; used to refer to circumstances in which toxicity is possible under normal conditions of use
7. microorganisms or substances capable of producing disease	9. a measure of the probability and severity of doing harm
10. whatever remains; in the case of pesticides, those amounts that remain on foods when people buy and use them	

⑥ Chapter 19 Answer Key ⑨

Summing Up

1. foodborne
2. pesticides
3. intoxication
4. natural
5. cross-contamination
6. cold
7. washed
8. raw

9. irradiation
10. wash
11. short
12. minimal
13. alert
14. public
15. variety
16. crop

17. inappropriately
18. FDA
19. organic
20. Additive
21. environmental
22. pesticide
23. EPA

Chapter Study Questions

1. Foodborne illness is the leading food safety concern. An estimated 76 million people per year experience foodborne illness, and for 5,000 of them it can potentially be fatal.

2. Two types of foodborne illnesses include those caused by an infectious agent (foodborne infection)—for example, *Campylobacter jejuni*; and those caused by a poisonous substance (food intoxication)—for example, *Clostridium botulinum*. Measures to prevent them include: use proper canning methods; avoid commercially canned foods with leaky seals, or with bent, bulging or broken cans; cook food thoroughly; use sanitary food handling methods; avoid unpasteurized milk; avoid raw fruits and vegetables where protozoa are endemic; dispose of sewage properly; and refrigerate foods promptly and properly.

3. Wash all surfaces that have been in contact with raw meats, poultry, eggs, fish and shellfish before reusing; serve cooked meats, poultry and seafood on a clean plate. Separate raw meats and seafood from those that have been cooked. Do not use marinade that was in contact with raw meat. When cooking meats, use a thermometer to test the internal temperature, and cook to the temperature indicated for that particular meat. Cook hamburgers to at least medium well-done. Cook stuffing separately or stuff poultry just prior to cooking. Do not cook large cuts of meat or turkey in a microwave oven. Cook eggs before eating them. Cook seafood thoroughly. When serving meats and seafood, maintain temperature of 140 degrees or higher, and heat leftovers thoroughly to least 165 degrees.

4. Stubborn or enduring continuance; the quality of persisting, rather than breaking down, in the bodies of animals and human beings. Contaminants get into foods: heavy metals and other contaminants entering the air in smokestack emissions return to the soil in rainfall, contaminants in the soil are absorbed by plants. People either eat the plants (fruits and vegetables) or meat from livestock that have eaten the plants. Sewage sludge and pesticides leave residues in the soil; runoff pollutes ground and surface water and contaminates the seafood that people eat. Toxins in the food chain accumulate: a person whose principal animal-protein source is fish may consume about 100 pounds of fish in a year, and these fish will have consumed a few tons of plant-eating fish in the course of their lifetimes; the plant eaters will have consumed several tons of photosynthetic producer organisms. If the producer organisms have become contaminated with toxic chemicals, these chemicals become more concentrated in the bodies of the fish that consume them. If none of the chemicals are lost along the way, one person ultimately eats the same amount of contaminant as was present in the original several tons of producer organisms.

5. Poisonous mushrooms are natural yet can be dangerous when eaten. Cabbage, turnips, mustard greens, and radishes contain small quantities of goitrogens that can enlarge the thyroid gland; this can cause problems if a person with a thyroid problem consumes large quantities of these foods. Lima beans and some fruit seeds contain cyanogens that, if activated, can produce the deadly poison cyanide. Potatoes contain small amounts of natural poisons. Poisons are poisons whether made by man or by nature.

6. Pesticides may become hazardous by remaining on crops, polluting water, contaminating the soil, and accumulating in the tissues of animals. The EPA determines what levels of pesticide residues in the food supply are considered acceptable. The FDA is responsible for enforcing adherence to this tolerance level, and monitors pesticide residues by collecting samples of foods and testing crops taken directly from the fields. Buy fresh foods grown locally, using responsible methods, and buy a variety of foods.

7. A GRAS substance (such as salt) is accepted as safe based on long experience consistent with the belief that it is not hazardous, whereas a regulated food additive (such as MSG) has been chemically tested to ensure its effectiveness and safety. Types of food additives include antimicrobial agents, which prevent growth of microorganisms; antioxidants, which delay/prevent oxidative damage (e.g. rancidity); colors and flavors, which enhance appearance and taste, respectively; emulsifiers and gums, which improve consistency by thickening or stabilizing the food; and nutrients (vitamins and minerals).

Sample Test Questions

1. c (p. 663)
2. a (p. 663)
3. d (p. 663)
4. c (p. 663)
5. a (p. 664)
6. c (p. 664)
7. b (p. 664)
8. e (p. 664)

9. d (p. 665)
10. a (p. 666)
11. d (p. 666)
12. b (p. 666)
13. b (p. 667)
14. e (p. 668)
15. d (p. 667)
16. a (p. 668)

17. d (p. 670)
18. c (p. 671)
19. a (p. 672)
20. a (p. 672)
21. b (p. 675)
22. e (pp. 664, 678)
23. b (p. 678)
24. b (p. 677)

25. c (p. 682)
26. b (p. 682)
27. a (p. 682)
28. d (p. 683)
29. b (p. 682)
30. c (p. 685)

Crossword Puzzle

1. persistence
2. certification
3. tolerance level

4. solanine
5. irradiation
6. safety

7. pathogens
8. hazard
9. risk

10. residues

⑤ Chapter 20 – Hunger and the Global Environment ⑥

Chapter Outline

I.   Hunger in the United States
   A. Defining Hunger in the United States
   B. Relieving Hunger in the United States
      1. Federal Food Assistance Programs
      2. National Food Recovery Programs
      3. Community Efforts
II. World Hunger
   A. Food Shortages
      1. Political Turbulence
      2. Armed Conflicts
      3. Natural Disasters
   B. Malnutrition
   C. Diminishing Food Supply
III. Poverty and Overpopulation
   A. Population Growth Leads to Hunger and Poverty
   B. Hunger and Poverty Lead to Population Growth
   C. Breaking the Cycle
IV. Environmental Degradation and Hunger
   A. Environmental Limitations in Food Production
   B. Other Limitations in Food Production
V.   Solutions

   A. Sustainable Development Worldwide
   B. Activism and Simpler Lifestyles at Home
      1. Government Action
      2. Business Involvement
      3. Education
      4. Foodservice Efforts
      5. Individual Choices
VI. Progress toward Sustainable Food Production
   A. Costs of Producing Food Unsustainably
      1. Resource Waste and Pollution
         a. Planting Crops
         b. Raising Livestock
         c. Fishing
         d. Energy Overuse
         e. The Cumulative Effects
      2. From Family Farms to Agribusiness
   B. Proposed Solutions
      1. Low-Input Agriculture
      2. Precision Agriculture
      3. Agricultural Biotechnology
      4. Consumer Choices
         a. Plant versus Animal
         b. Local versus Global

Summing Up

    Consumers grow curious about the (1)_____ impacts of their food choices. Alternative food choices that are more environmentally benign are now (2)_____. This chapter emphasizes personal lifestyle (3)_____ because they raise awareness of the need for larger actions and (4)_____ the way for them.

    Many trends are occurring that contribute to global (5)_____ that are related in that their (6)_____ overlap as well as their (7)_____. Environmentally conscious food (8)_____, preparing, cooking, and transporting can assist in solving related problems. Using proper cooking (9)_____ and kitchen appliances can also be beneficial.

    This chapter also addresses the problems of (10)_____ and poverty in the United States and (11)_____ countries. These problems have always existed and despite numerous (12)_____, the number of hungry and poor people continues to grow.

    Many nations now recognize that improvement of all nations' economies is a prerequisite to meeting the world's other urgent needs: (13)_____ stabilization, arrest of environmental (14)_____, sustainable treatment of (15)_____, and relief of hunger.

Chapter Study Questions

1. Identify some reasons why hunger is present in a country as wealthy as the United States.

2. Identify some reasons why hunger is present in the developing countries of the world.

3. Explain why relieving environmental problems will also help to alleviate hunger and poverty.

4. Discuss the different paths by which rich and poor countries can attack the problems of world hunger and the environment.

5. Describe some strategies that consumers can use to minimize negative environmental impacts when shopping for food, preparing meals, and disposing of garbage.

Chapter Glossary

- **emergency shelters:** facilities that are used to provide temporary housing.
- **famine:** extreme scarcity of food in an area that causes starvation and death in a large portion of the population.
- **field gleaning:** collecting crops from fields that either have already been harvested or are not profitable to harvest.
- **food bank:** a facility that collects and distributes food donations to authorized organizations feeding the hungry.
- **food insecurity:** limited or uncertain access to foods of sufficient quality or quantity to sustain a healthy and active life.
- **food insufficiency:** an inadequate amount of food due to a lack of resources.
- **food pantries:** programs that provide groceries to be prepared and eaten at home.
- **food poverty:** hunger resulting from inadequate access to available food for various reasons, including inadequate resources, political obstacles, social disruptions, poor weather conditions, and lack of transportation.
- **food recovery:** collecting wholesome food for distribution to low-income people who are hungry.
- **food security:** certain access to enough food for all people at all times to sustain a healthy and active life.
- **fossil fuels:** coal, oil, and natural gas.
- **human carrying capacity:** the maximum number of people the earth can support over time.
- **nonperishable food collection:** collecting processed foods from wholesalers and markets.

- **oral rehydration therapy (ORT):** the administration of a simple solution of sugar, salt, and water, taken by mouth, to treat dehydration caused by diarrhea.
- **perishable food rescue** or **salvage:** collecting perishable produce from wholesalers and markets.
- **prepared food rescue:** collecting prepared foods from commercial kitchens.
- **soup kitchens:** programs that provide prepared meals to be eaten on site.
- **sustainable:** able to continue indefinitely; using resources at such a rate that the earth can keep on replacing them and producing pollutants at a rate with which the environment and human cleanup efforts can keep pace, so that no net accumulation of pollution occurs.

Sample Test Questions

Select the best answer for each question.

1. An estimated 1 out of every _____ people worldwide live in poverty:
   a. 4
   b. 6
   c. 8
   d. 10
   e. 20

2. Certain access to enough food for people to sustain a healthy and active life at all times is:
   a. food recovery.
   b. food security.
   c. food management.
   d. food access.

3. Which of the following statements is true regarding food availability?
   a. Due to agricultural bounty in the U.S., hunger is not a problem.
   b. Due to enormous wealth in the U.S., food insecurity is not an issue.
   c. Over 38 million people in the U.S. live in poverty and cannot afford to buy enough food.
   d. One out of 30 households in the U.S. experience hunger or the threat of hunger.

4. The limited or uncertain access to foods of sufficient quality or quantity to sustain a healthy life is:
   a. food poverty.
   b. hunger.
   c. food recovery.
   d. food insecurity.

5. People who have too little food and try to stretch their limited resources by eating small meals or skipping meals are experiencing:
   a. hunger.
   b. food insufficiency.
   c. food poverty.
   d. famine.

6. Reasons for food poverty include:
   a. illnesses and disabilities.
   b. unemployment, low-paying jobs, and medical expenses.
   c. alcohol and other drugs.
   d. a and b
   e. a, b and c

7. Which of the following statements is true about hunger?
   a. People who are hungry are underweight.
   b. The highest rates of obesity occur among those living in the greatest poverty.
   c. People who are obese have plenty of income to purchase large amounts of food.
   d. Food insecure people who do not participate in food assistance programs have lower rates of obesity.

8. The largest food assistance program is:
   a. WIC.
   b. the School Lunch Program.
   c. Meals on Wheels.
   d. the Food Stamp Program.

9. Collecting wholesome food for distribution to low-income people who are hungry is:
   a. food recovery.
   b. food security.
   c. food banking.
   d. food pantries.

10. Common methods of food recovery include:
    a. field gleaning and perishable food rescue.
    b. prepared food rescue and nonperishable food collection.
    c. dumpster food rescue and soup kitchens.
    d. a and b
    e. a, b and c

11. Collecting crops from fields that have already been harvested is specifically called:
    a. food insecurity.
    b. field gleaning.
    c. food rescue.
    d. food collection.

12. Extreme scarcity of food that causes starvation is:
    a. extinction.
    b. deforestation.
    c. sustainable.
    d. famine.
    e. poverty.

13. To help prevent measles mortality and blindness, _____ are distributed to children worldwide.
    a. large containers of rice
    b. bottles of clean water
    c. vitamin A supplements
    d. bottles of Ensure

14. Administration of a sugar, salt and water solution is called:
    a. ORT.
    b. FDA.
    c. WHO.
    d. ADA.

15. The primary cause of hunger is:
    a. lack of education about food selection and preparation.
    b. abuse of alcohol and other drugs.
    c. depression.
    d. poverty.
    e. mental illness.

16. Communities can address their hunger problems by:
    a. improving public transportation to food resources.
    b. integrating public and private resources to relieve hunger.
    c. identifying high-risk populations and target services to meet their needs.
    d. all of the above

17. The maximum number of people the earth can support over time is its:
    a. human carrying capacity.
    b. overpopulation threshold.
    c. population demand.
    d. sustainability index.

18. Which of the following statements is true regarding poverty and overpopulation?
    a. Women living in poverty have fewer children than women with financial means.
    b. Women in poverty are treated well by men.
    c. In some areas, children raised in poverty are economic assets.
    d. Families living in poverty choose to live the way they do.

19. Environmental problems that slow food production include:
    a. soil erosion, compaction, and salinization.
    b. deforestation and desertification.
    c. air pollution, ozone depletion and climate changes.
    d. water pollution and scarcity.
    e. all of the above

20. Using resources at a rate at which the earth can continue replacing them is called:
    a. degradation.
    b. sustainable use.
    c. population growth.
    d. carrying capacity.

21. Urgent needs for all nations' economies include:
    a. hunger relief and population stabilization.
    b. environmental preservation and sustainable resources.
    c. nutrition education and food giveaways.
    d. a and b
    e. a, b and c

22. Selected segments of society that can make the greatest impact in fighting hunger, poverty and environmental degradation include:
    a. federal and state governments.
    b. companies and educators.
    c. individuals.
    d. a and b
    e. a, b and c

23. Fossil fuels include:
    a. ozone.
    b. coal.
    c. oil.
    d. a and b
    e. b and c

24. If a person responds positively to the question, "Our children are not eating enough because we couldn't afford food," this indicates:
    a. food withholding.
    b. food insecurity.
    c. poor parenting.
    d. food security.

25. To save money and spend wisely, follow this rule when shopping for food:
    a. shop when hungry.
    b. buy large amounts of meat.
    c. purchase oatmeal instead of ready-to-eat cereals.
    d. buy name brands only.

26. To make environmentally friendly food-related choices:
    a. walk, bike ride or use mass transit.
    b. shop less often.
    c. when shopping for a car, choose an energy-efficient model.
    d. all of the above

27. When selecting foods low on the food chain:
    a. you are consuming more fish products.
    b. you are eating better for your health.
    c. you are harming the environment.
    d. you are violating the *Dietary Guidelines*.

28. Environmentally-friendly purchasing includes:
    a. selecting local foods.
    b. purchasing products at large national chains.
    c. purchasing foods with packaging.
    d. purchasing eggs in foam containers.

29. Environmentally friendly cooking tips include:
    a. use a microwave oven.
    b. use a convection oven.
    c. purchase only raw foods.
    d. all of the above

30. To be environmentally-friendly, one should:
    a. not use disposable plates and cups.
    b. use cloth towels and napkins.
    c. run the dishwasher only when it is full.
    d. recycle glass, plastic and aluminum.
    e. all of the above

Crossword Puzzle

Complete this crossword puzzle by Mary A. Wyandt, Ph.D., CHES.

Across:	Down:
1. a common method of food recovery; _____ *food rescue*	2. programs that provide groceries to be eaten on site; often called soup kitchens
4. coal, oil, and natural gas; these are nonrenewable fuels that pollute	3. a common method of food recovery known as _____ or *perishable food rescue*
5. _____ or *alternative fuels*, such as solar and wind energy, pollute less or not at all	4. a common method of food recovery known as _____ *gleaning*
7. an inadequate amount of food due to a lack of resources	6. the maximum number of people the earth can support over time is called the *human* _____ *capacity*
9. a white, fluffy cash crop	8. *(abbrv.)* the administration of a simple solution of sugar, salt and water, taken by mouth, to treat dehydration caused by diarrhea

⑥ Chapter 20 Answer Key ⑥

Summing Up

1. environmental
2. available
3. choices
4. pave
5. problems
6. causes
7. solutions
8. shopping
9. methods
10. hunger
11. developing
12. programs
13. population
14. degradation
15. resources

Chapter Study Questions

1. Hunger is present in the U.S. because of the many political, social, and economic factors related to poverty. The primary cause of hunger in the U.S. and other developed countries is food poverty; i.e., people live in hunger because their income is too low to purchase enough food.
2. Poverty, famine due to political unrest, armed conflict, or natural disasters, and a diminishing food supply.
3. Relieving environmental problems can help eliminate hunger and poverty in several ways: by reducing soil erosion, the agricultural productivity can be improved; decreasing deforestation will decrease soil erosion; and reducing air pollution will increase crop yields. Reducing water pollution can improve agricultural and fishery production.
4. The poor nations need to reduce population growth; reverse destruction of forests, waterways, and soil; and reduce poverty. Rich nations need to reduce wasteful and polluting uses of resources and energy; and relieve debtor nations of poverty.
5. Shoppers can car pool, use mass transit, walk or bicycle to shop, shop less often—make fewer trips, or take turns shopping for each other. When preparing foods: cook mostly plant foods, use a pressure cooker and microwave or stir-fry foods, use the oven and stove top less often, and do without small electrical appliances. When disposing of garbage: find uses for items normally thrown away, recycle trash, and refuse to purchase items that have excess packaging.

Sample Test Questions

1. c (p. 702)
2. b (p. 702)
3. c (p. 702)
4. d (p. 702)
5. b (p. 702)
6. e (p. 702)
7. b (p. 702)
8. d (p. 703-704)
9. a (p. 704)
10. d (p. 704)
11. b (p. 704)
12. d (p. 705)
13. c (p. 706)
14. a (p. 707)
15. d (p. 708)
16. d (p. 711)
17. a (p. 707-708)
18. c (p. 708)
19. e (p. 709)
20. b (p. 711)
21. d (p. 710-711)
22. e (p. 711)
23. e (p. 709)
24. b (p. 702)
25. c (p. 704)
26. d (p. 712)
27. b (p. 712)
28. a (p. 712)
29. a (p. 712)
30. e (p. 712)

Crossword Puzzle

1. prepared
2. emergency kitchen
3. salvage
4A. fossil fuels
4D. field
5. renewable
6. carrying
7. food insufficiency
8. ORT
9. cotton